Groupe International de Recherche sur l'Enseignement de la Physique Seminar (GIREP 2016)

Research-Based Proposals for Improving Physics Teaching and Learning – Focus on Laboratory Work

Journal of Physics: Conference Series
Volume 1076

Krakow, Poland
30 August - 3 September 2016

Editor:

Dagmara Sokolowska

ISBN: 978-1-5108-7511-1
ISSN: 1742-6588

Printed from e-media with permission by:

Curran Associates, Inc.
57 Morehouse Lane
Red Hook, NY 12571

Some format issues inherent in the e-media version may also appear in this print version.

Printed by Curran Associates, Inc. (2018)

For permission requests, please contact the Institute of Physics
at the address below.

Institute of Physics
Dirac House, Temple Back
Bristol BS1 6BE UK

Phone: 44 1 17 929 7481
Fax: 44 1 17 920 0979

techtracking@iop.org

Additional copies of this publication are available from:

Curran Associates, Inc.
57 Morehouse Lane
Red Hook, NY 12571 USA
Phone: 845-758-0400
Fax: 845-758-2633
Email: curran@proceedings.com
Web: www.proceedings.com

Table of contents

Volume 1076

GIREP SEMINAR 2016 - 30 August to 3 September 2016, Kraków, Poland

Accepted papers received: 25 July 2018

Preface

Papers

PART I: GENERAL ISSUES

PART II: EXPERIMENTAL LAB IN INTRODUCTORY PHYSICS COURSES

PART III: MODERN PHYSICS AND ADVANCED LABS

PART IV: LAB WORK AND MULTIMEDIA

IOP Conf. Series: Journal of Physics: Conf. Series **1076** (2018) 011001 doi:10.1088/1742-6596/1076/1/011001

Groupe International de Recherche sur l'Enseignement de la Physique

GIREP SEMINAR 2016
August 30 – September 3, 2016
Kraków, Poland

50th Anniversary GIREP

Research-based proposals for improving physics teaching and learning – focus on laboratory work

Organized by:
International Research Group on Physics Teaching (GIREP)
Faculty of Physics, Astronomy and Applied Computer Science
of the Jagiellonian University in Kraków

Editor
Dagmara Sokołowska

Editorial Secretariat
Anna Gagatek

IOP Conf. Series: Journal of Physics: Conf. Series **1076** (2018) 011001 doi:10.1088/1742-6596/1076/1/011001

Honorary patronage

We greatly appreciate the honorary patronage of:

Prof. dr hab. Andrzej Białas
President of the Polish Academy of Arts and Sciences (PAU)

Jarosław Gowin
Minister of Science and Higher Education

We would like to acknowledge the financial support from KNOW grant no. 62/KONF/DS/2016.

IOP Conf. Series: Journal of Physics: Conf. Series **1076** (2018) 011001 doi:10.1088/1742-6596/1076/1/011001

Preface

The Proceedings present selected contributions from the international conference GIREP Seminar 2016, organized by GIREP vzw organization and the Faculty of Physics, Astronomy and Applied Computer Science at the Jagiellonian University, Kraków, Poland. It was our great privilege to host GIREP members and friends in the year of the 50th Anniversary of GIREP organization. The first day of this event offered an opportunity to recall special memories and to thank everyone that has contributed to the growth of GIREP during the last half-century.

The general seminar topic **Research-based proposals for improving physics teaching and learning – focus on laboratory work** emphasized the importance of laboratory activities in physics education. The overall aim of this seminar was to highlight the various aspects of laboratory work involved in establishing an environment where physics teaching and learning can take place, and in particular the development of physics literacy. Several topics have been discussed in order to line out a wider view of laboratory work at all levels of physics and science education, from primary school to physics courses at the university.

The format of this seminar was proposed in the style of the old-time GIREP meetings – with keynotes, oral presentations and poster presentations focused on six themes, followed by in-depth discussions in small groups of researchers and practitioners in sessions led by leaders of six Working Groups (WG). The contributions from six keynote speakers, widely respected in the community of physics education, as well as a comprehensive variety of oral and poster contributions, offered an unforgettable occasion for a fruitful exchange of thoughts and ideas.

The impact of physics education research on the educational design and practice of physics laboratory was the focus of WG1: *Experimental Lab in Introductory Physics Courses.* Presentations showed studies of students' learning in the laboratory and difficulties they come across, as well as, teaching proposals for specific topics at secondary schools, colleges and the first years of university. In WG2 two topics were encompassed. *Advanced Experimental Laboratories*, rarely addressed by instructors and researchers, who are focused more on introductory physics labs, was chosen to fill this gap and open a broader discussion on the role, goals and examples of the advanced laboratories in physics student education during their bachelor and master studies. *Modern Physics* topics being of the most interest of learners at all ages, appear to be rarely addressed in high school and during the first years of physics studies due to time limitations and the lack of teachers' competences. Contributors taking part in discussion tried to answer the question how to translate complex theories and highly-advanced experiments into language understandable and appreciated by less advanced students.

Since a modern laboratory can barely be operated without ICT, thus the design, evaluation and characterization of resources and environments for physics teaching and learning with use of ICT was addressed in WG3: Lab Work and Multimedia. Participants focused in particular on online learning environments, simulation and modeling tools, virtual laboratories and open sources. Self-regulation, reflection and collaboration in digital learning environments in context of lab work were discussed. WG4 *Conceptual Lab and Mathematization* addressed theories, models, and empirical results on conceptual understanding, conceptual change and development of competences in context of laboratory work, as well as methodology for investigating students' processes of concept formation and concept use on the basis of experiments and strategies to promote conceptual development throughout laboratory activities. A broad meaning of the term *mathematics* that includes all kinds of structuring and ordering physical processes: using abstract methods like idealization and modeling, as well as using a broad range of mathematical elements such as diagrams, graphs and formalized sketches (e. g. arrows) and equations was discussed in the context of physics laboratories.

A specific role and character of laboratory activities encourage the teachers to search for non-standard assessment strategies. In lab more than in other physics learning environments the formative assessment for development of research skills and conceptual understanding plays a dominant role. A detailed discussion on that topic was the core of WG5: *Assessment for learning through experimentation*.

In order to attract more students, attention for science should be brought naturally, with use of everyday materials and in everyday context. Understanding of physics and appreciation of its beauty starts when observing usual but at the same time – amazing phenomena around. Traditional laboratory environment is extended nowadays beyond the lab space. Experiments are shown and tried out during numerous shows, festivals and other experiences outside the classroom, including those organized by institutions other than schools. Simple experiments should serve as the ignition of ideas, concepts and the notion for development of intuition in physics, not only at early ages, but across entire education. There aspects of learning, additional to traditional education, were the point of discussion in WG6: *Low Cost Experiments and Inquiry*.

The seminar was attended by 115 participants representing 28 countries. The scientific seminar program offered altogether 6 invited talks, 63 oral and 42 poster presentations. After the seminar 54 papers were received on all seminar topics. The articles went through a rigorous process of in a double-blinded peer-review, involving members of the Editorial Board and twelve additional referees in order to guarantee the quality of the content of this contribution. As the result two publications are issued, the book *Focusing on Lab to improve Physics Teaching and Learning. Research Based Proposals*, published by Springer and GIREP Seminar 2016 Proceedings book, presented here.

The organization of the seminar would not have been possible without help and co-operation of many people. First of all, we would like to thank Prof. Marisa Michelini, GIREP President, for her constant help and support. We sincerely thank the members of the Advisory Board and colleagues on the Local Organizing Committee for their dedication and commitment to this event. We are also deeply thankful to all reviewers, Working Group Leaders and the Head of all Leaders, Dr. Ian Lawrence. We would like to express our gratitude to six invited speakers for their valuable presentations that served as the foundation for the group discussions throughout the entire seminar. We are also deeply indebted to Prof. Paul Black who joined the anniversary day of GIREP with his special talk on-line and to Dr. Seta Oblak and Dr. Zofia Golab-Meyer for their contribution to the seminar on the history of GIREP and its impact on physics education research and development.

We would also like to thank all the participants of the GIREP Seminar 2016, for submitting proposals, advance preparations for discussions and sharing their ideas with the GIREP community. We hope that these Proceedings will give the reader an opportunity for deeper comprehension of the Laboratory Work aspects to improve physics teaching and learning.

Kraków, Poland

Dagmara Sokołowska

IOP Conf. Series: Journal of Physics: Conf. Series **1076** (2018) 011002 doi:10.1088/1742-6596/1076/1/011002

Peer review statement

All papers published in this volume of *Journal of Physics: Conference Series* have been peer reviewed through processes administered by the proceedings Editors. Reviews were conducted by expert referees to the professional and scientific standards expected of a proceedings journal published by IOP Publishing.

IOP Conf. Series: Journal of Physics: Conf. Series **1076** (2018) 012001 doi:10.1088/1742-6596/1076/1/012001

Teaching physics in XXI century. Why and how

Łukasz Turski

Center for Theoretical Physics of the Polish Academy of Sciences, Warsaw, Poland

Fifty years ago, when GIREP was founded, our civilization was very different from that we are now living in and which we are trying to comprehend in order to provide at least some guidance to the future generations. The guidance which in the past we would have called education.

Fifty years ago our civilization was that of vinyl records, Walkman and MP3 era was yet to come, electronic watches and wireless phones were only in the Dick Tracy cartoons and the James Bond movies. Computers were behemoths hidden behind doors of military, industrial and selected research institutions. CERN an Arecibo were in their infancy and so were preparations to the moon landing. Satellites were launched but they have served mostly military purposes with very limited though often important scientific output.

Fifty years ago the world was heavily breathing under the weight of the Cold War, regaining semi-balance after the President Kennedy assassination and the Chairman Khrushtschev forced resignation. France was struggling with relocation of almost 900 thousands of "migrants" from Algeria and mopping out the rest of OAS. The new seeds of terror cancer were growing in Germany and Italy. The sounds of guns from Vietnam were not yet bothering us too much.

In spite of all that "the West" was rich and happy but soon this state of illusionary tranquility was shattered on streets of Paris with student revolt which had derailed education system of those times. The system which was essentially the continuation of "the business as usual" inherited from the pre-war Europe. Major consequence of that Paris spring was that the era of experiments in education based upon preconceived ideological "principles" has begun. The best description of those attempts to invent new education was ridiculed by comedian and mathematician Tom Lehrer in his song "New Math"[1] – "the important thing is to know what are you doing rather than to get right answer". Politically motivated system of tests was slowly but persistently taking over, first schools and later, universities. That was, basically the educational systems which XXI century generations have inherited.

The beginning of XXI century was marked by the first sign that our society is not prepared to face reality of the technological changes, which throughout the last years of the previous century were already slowly and quietly reshaping the future of the mankind. The bogus Y2K catastrophe, predicted for the 2000-2001 night, was quickly forgotten and the lesson from it has never been learned. Meanwhile, progressing like the bushfire, the process of moving all our essential activities *on-line* resulted in the situation that serious collapse of the information network in any country, but especially in the most developed ones, would be an effective Armageddon.

On various occasions I have been showing the picture, taken by my nephew, during the New York hurricane Sandy blackout in 2012, of people on the street lining up to the stand providing the charge to their mobile devices. Access to the Net was equally important as food or water.

Today's family, with two junior-high age kids, is typically using four smartphones. A rather conservative estimate of yearly electric energy share of the smartphone, in support of the network and world cloud system – we can neglect minor energy use for charging – is roughly the same as that of

[1] Tom Lehrer. New Math. https://www.youtube.com/watch?v=UIKGV2cTgqA.

a modern refrigerator. That family is surely completely unaware of that although it will be unable to exist, socially and professionally, without access to the Net. Imagine what will happen when, eventually, someone will ask them to pay the bill? In spite of taking hours of physics in high schools and possibly in college they were never properly thought to understand the energy flow in the contemporary society.

In that completely different civilization our educational system is trying to provide the guidance to our youth using essentially the same language and ideas as fifty years ago. We are as physicists on the turn of XIX and XX century trying to explain various, discovered at that time phenomena with classical physics. They failed since what was needed was a different science and a different language. Eventually that new science – quantum mechanics and relativity theory were established changing world forever. That is precisely what we need now in education.

Since the civilization, we lived in, was changed by the proliferation of the discoveries of physics, we are obliged to start the invention of new education, education of the XXI century, from changing the way we teach physics. Physics is understood here in a broad sense as what used to be called natural sciences, mathematics and applications in some basic social sciences like economy. That is the definition of physics given in the seminal lecture of great mathematician Vladimir Arnold back in the XX century[2].

Physics is today the only science which amalgamates qualitative and quantitative description of phenomena ranging from the properties of vacuum to those of the whole universe. It does it using remarkable simple set of fundamental rules and employing particular way of solving problems. That is by reduction of a problem to few essential elements which either were already solved or, when not, were then experimentally investigated to provide grounds for building theoretical model. That way of understanding phenomena is what is needed in copying effectively with the current civilization transformation. Other attempts result in not rational but chaotic activities generating more entropy than it is required by the Second Law.

We should base our future educational system, or systems, on problem solving. The problems are all around us. Thomas Jefferson formulated the basic ideas of science and civilization progress, which Gerald Holton called Jeffersonian Research Program[3]. It assumes that problems to be solved are picked by active individuals depending on his/her interest from the pool of issues provided by natural and social, cultural and political environment. That program, applied to education, reverberates XVIII century Johann Pestalotzzi "teaching child not a subject" program. We should allow a pupil to pick up problems according to his or her talents and interest and then use that knowledge wisely to steer that person throughout the long process of education. That requires individualization of education, not possible in the past but achievable now due to phenomenal development of technology. We no longer need to provide the same chunk of knowledge to everybody, for whatever knowledge of facts, data, methods, is necessary it is available in that sea of information provided by the global network. What we have to provide in education is how that knowledge should be effectively filtered out from the peta-bytes of information and noise of the Network. Because effective filtering of information is exactly what we do in physics and since most of the problems of today are physics-related that is why we need to teach physics in XXI century. The question is now how.

Since physics is to serve as a tool of education rather than just one of many subjects included in the school curricula the way we teach it should change. I have already mentioned conservation of energy. Probably every program of high school physics contains exposition to the conservation of energy and basis of thermodynamics. In spite of that, the concept "renewable energy" is used permanently in the important debate about the mankind energy and/or climate future. The fact that in Europe solar energy per square meter per day only implies 0.5 l of 95 octane gasoline and that this tiny amount is further

[2] V. Arnold. On Teaching Mathematics cf for example http://pauli.uni-muenster.de/~munsteg/arnold.html.

[3] Gerald Holton. Science and Anti-Science Harvard University Press, Boston 1993 see also http://issues.org/16-1/holton/.

IOP Conf. Series: Journal of Physics: Conf. Series **1076** (2018) 012001 doi:10.1088/1742-6596/1076/1/012001

partially lost by whatever devices we invent to use it, because thermodynamics laws are so "cruel" is not generally understood, not only by politicians.

A brief check of the Tesla cars add page[4] provide information, given by the producer, on how much electricity we need to produce and to deliver to the electric outlet in the Tesla owner garage in order to provide the juice for a one kilometer ride. Multiplying that by the number of cars registered in a country and then by averaged mileage covered we can easily estimate that switching to Teslas would require increase of electricity production of a country by huge percentage (in Poland about 15 to 30%). That would be a disaster to a country electric grid, unless it will be considerably improved.

We cannot prepare the future generations for the „all electric" world unless we will teach them science, particularly physics, differently than today. The founding father of the electric revolution of previous century, Carl Proteus Steinmetz, was aware of that when he was preparing courses for his students in Union College. His ideas echo in words of Thomas Friedman and Michael Mandelbaum[5].

> *Learning to repair the engine of an electric car, or a robotic cutting tool, or a new gas-powered vehicle that has more computing power that the Apollo space capsule – these are no skills you can pick up in a semester of high school class.*

Staying with my example of the science of electricity, that implies that in addition to fundamental facts like Coulomb interaction, Kirchhoff's and Ohm's laws we have finally to include into high-school curricula Maxwell equations. For in the world which is now immersed in the electromagnetic radiation, form the sun light to Bluetooth generated waves connecting all the smartphones, laptops etc. in this lecture hall, knowledge of the properties of the electromagnetic waves is of predominant importance.

To teach the properties of the electromagnetic fields and their interaction with matter we also have to teach how the mankind progressed by developing the means of generating electric energy including the fact that electricity is a fantastic tool to transport energy and extremely ineffective way of storing it. The knowledge of that is required for all citizens of XXI century who have make a decision on how in the future we will generate electric energy. Otherwise various activists will continue to win debates for and against nuclear energy using solar energy as panacea for our energetic dilemmas. The modern teaching of electricity and magnetism will have to abandon beautiful and pretty useless, for today discussion, XIX century experiments, they have to be replaced by modern experiments showing all the same properties of the electromagnetic field. The ghost of a one Farad, a giant metal sphere, should be either replaced with the capacitor Earth ionosphere – ground or be buried with all required honors. I would like to ask this conference how many students, you teach, know how the super-capacitor, in my few dollars' worth watch, works. Just the basic principle.

The return of physics teaching to its usefulness begins in the school laboratory. Only by giving students the chance of doing lots of modern experiments we will be able to show them, that it is impossible to do those experiments and subsequently build "things" we use daily, without thorough knowledge of the fundamental laws, like the conservation of energy I was just talking about.

We should also renew the pool of experiments we do in schools. We have to stop fearing experiment with radioactivity in schools. Knowledge of radioactivity is an anathema in today's curricula, an absurd in the time when our medical colleagues are using antimatter in every day practice (PET devices). Irresponsible choice of the radioactivity units result in the fact that seven or so kilo-becquerels of radioactive decays in our body and even the radioactivity of common banana might stir a profound feeling of fear between most of the college educated individuals. How many radiologist attendants can explain the crucial difference between the Gray and Sievert units. That difference is important in explaining why beaches of Brazilian resort Guarapari are open to the public.

---

[4] https://www.tesla.com.

[5] T.L. Friedman & M. Mandelbaum. That Used to Be Us How America Fell Behind in the World It Invented and How We Can Come Back Picador New York (Kindle edition).

There are many areas of classical physics, whatever that means in the XXI century, which are wrongly presented in schools. For example the notion of entropy is so crucial in understanding contemporary informatics, how the MP3 or other compressing algorithms work etc. I do not even mention the basis of quantum information processing. That last idea might revolutionize the world beyond the common tale about Schrödinger cat. Do we teach that at all?

Most of mechanical devices function because of friction. How many thousands of students are finishing their school or college education with the wrong „conviction" that friction force T is always equal of ηN, where N is the force exerted perpendicularly to the surface and η is the friction coefficient?

Essentially all the facts about things I have mentioned can be found in the Net. Most are there to find using ways of searching we have to explain to students, particularly explaining that what is on the Web is not necessarily true. Wikipedia, unfortunately, is not as dependable source of information as the Wolfram Alpha. Not all the e-education materials are of the quality of Khan Academy.

IT proliferation had changed our life. One of the area of education where, I believe, the use of the IT should be restricted, is the school laboratory. I strongly oppose the replacement of real experiments with the beautiful computer simulations and/or YT presentations. Computers can and should help us to do experiments better and analyze their results but should never replace the experiment. I repeat here again that the only use of laptop, tablet or whatever, in teaching gravity is to drop it from the table to show which direction the earth gravity works. A few dollars' worth tools available in the Raspberry Pi project, together with that computer by itself, are excellent in helping students to build their own experimental apparatus required to precisely measure the gravity acceleration. However, that would require an active participation of a student in a real experiment and offers them the chance to face difficulties of the real experiment–that means world, not beautiful computer graphics occasionally massaged for the purpose of better visualization as the laws of the mechanics used in shooting fighting scenes in the movie Matrix.

I believe knowledge of physics is important as a basic tool for understanding the world. Three years ago my10 years old grandson took with him for vacation, we were spending together, the book suggested by his teacher – *The Adventures of Tom Sawyer.* I noticed that he did whatever possible to avoid reading it. He confessed that he found that book dull. He was reading the paper edition of the book. I promptly provided an electronic version of it and we started to read it using the quality book-reader on one of our tablets. The book had „opened up". I found that the children, here in Poland, asked to read that book are unaware from where it comes , and what is the meaning of, the Samuel Clemens literary pseudonym – Mark Twain. So we used the ability of the reader to search the Web for notions marked in the text and we started to explore first the notion – Mississippi. Soon were investigating the details of that river hydrology, then the engineering of the famous Mississippi boats and ships. For example why those boats have the paddle-wheel on the back in contrast to the boats, say on the Vistula or Rhine. Soon the question of the positioning of the boilers on those boats became an issue of discussion and simple experiments we did with breakfast frankfurters, etc. We also had to resolve the problem of the book translation into Polish, particularly the question of some politically incorrect words used by Twain, which have been translated into Polish in a completely ridiculous fashion. As you see I used the book from the primary-school children bookshelf as the tool to teach several "subjects"[6]. A few days ago, during the yearly conference for teachers, that was held at the Copernicus Science Center, jointly with the one of the leading Polish and French actor Andrzej Seweryn, we run the workshop devoted to the physics way of reading books. This time we used one of the rhymes for children by Polish poet Jan Brzechwa "The Nut" in which a lawyer from a little town next to Warsaw gets a particularly tough nut and tries to crush it by many means which all fail, eventually a squirrel

[6] L.A. Turski Lecture on the 60 anniversary of the Institute of Literature of the Polish Academy of Sciences 2013. Computer presentation available from the author. Fragments included in the interview with Anita Czupryn in *Polska the Times*, http://www.polskatimes.pl/artykul/1074970,prof-turski-bez-pana-tadeusza-nie-mozna-zrozumiec-tego-co-sie-w-polsce-dzialo-i-dzieje,id,t.html.

IOP Conf. Series: Journal of Physics: Conf. Series **1076** (2018) 012001 doi:10.1088/1742-6596/1076/1/012001

enters through the window and crushes the nut shell and eats the kernel. All those unsuccessful attempts to crush the nuts can be used to introduce the meaning of the word hard, the method of measuring hardness of materials (one of the attempts to crush the nut by a blacksmith using the heavy hammer fits beautifully the Leeb method of measuring the hardness by rebounding a tester head from a tested material) and finally the fascinating story of rodent's bite power. That last issue allows to introduce the concept of allometric relation, basic in nature but not mentioned in school education (the strength of rodent bite is related to the body mass of rodent by power law with the same exponent for hundreds of rodents investigated).

I believe I have made my points pretty clear. I would like to close with one quotation to support my way of thinking about the physics education of XXI century which is that from the book which has been my guidance in thinking about the education from the day I have got it, John Dewey *Democracy and Education: an Introduction to the philosophy of education*. It says:

> *Pupils begin their study of science with texts in which the subject is organized into topics according to the order of the specialist. Technical concepts, with their definitions, are introduced at the outset. Laws are introduced at a very early stage, with at best few indications of the way in which they were arrived at. The pupils learn „science" instead of learning the scientific way of treating the familiar material of ordinary experience.*

After fifty years of GIREP existence the world civilization is on a turning point. We have a fantastic progress in the technology, medicine, agriculture, biology–including that of our own genome structure and have possibility of making this world a place of plenty and as peaceful as possible. Nevertheless we are facing the migrants problem, on the scale similar to that fifty years ago , which we cannot handle with our technical means surpassing anything available to our predecessors. We are facing the terrorist treat which again we cannot contain in spite of employing technology no one even dreamed about when Red Brigades were rooming the streets of Rome. We face the world energy crisis and possible consequences of whatever changes in climate are ahead of us. I believe that all this is strongly related to the failure of the world educational system.

That Education system we have now failed us. The point is to build a new one. In that new system the main goal should be a better understanding of nature surrounding us, for the changes in it, irrespectively anthropogenic or caused by natural phenomena, will soon influence the way we live on a scale we had not envisage when the atomic energy was harnessed and transistors were build. This goal cannot be achieved without improving the understanding of basic rules the nature works according to and the basic rules through which we can use the nature in a sensible way. And that means physics. With improved physics education we can create truly educated society and the society of educated people, as Thomas Jefferson said, is the only one which can guarantee the preservation of the most precious value in human live-the freedom.

IOP Conf. Series: Journal of Physics: Conf. Series **1076** (2018) 012002 doi:10.1088/1742-6596/1076/1/012002

Understanding kinematics graphs using MBL tools, simulations and graph samples in an interactive engagement context in a Ghanaian university

Victor Antwi[1], Elwin Savelsbergh[2] and Harrie Eijkelhof[2]

[1]Department of Physics Education, University of Education, Winneba, Ghana
[2]Freudenthal Institute for Science and Mathematics Education, Utrecht University, the Netherlands

Abstract. Many students have problems when it comes to describing the shapes of displacement, velocity and acceleration-time graphs (x, v, a-t graphs), conversion of graphs from one form into another, and calculating and getting the meaning of slopes and areas under kinematics graphs. They often describe shapes of graphs as pictures and give interpretation without taking into cognizance of the type of graph being considered. In this study, 37 first year university physics students (Group One: 17 students and Group Two: 20 students) at the University of Education in Winneba (Ghana) in two consecutive years, were introduced to the use of microcomputer based laboratory (MBL) tools; simulations and graph samples to practice and describe the shapes of kinematics graphs; conversion of graphs from one form to the other; calculation of slopes and areas under kinematics graphs, and their meanings, all in an interactive engagement teaching. Students were made to answer the "Test of Understanding Graphs in Kinematics" (TUG-K) before and after the introduction of the use of MBL tools, simulations and graph samples. Students' scores were compiled and converted to mean proportion scores and average normalized gain $\langle g \rangle$, under the four concepts "Area under the graph (meaning and calculation); Slope (meaning and calculation); Graph description; and Graph transformation". The results indicate that the first year university students in the two groups all did better in describing the shapes of kinematics graphs, transforming kinematics graphs, calculating and getting the meaning of slopes and areas under kinematics graphs when they were tested with the same instrument after instruction in kinematics. This goes to show that MBL tools, simulations and graph samples when used in an interactive engagement manner can improve the teaching and learning of kinematics graphs in physics.

1. Introduction

Do you know that some students did not believe that graphs were representation of various kinds of quantitative information and relationships, such that their own movement (walking) could be plotted as graphs? To such students they see graphs as foreign materials, which have no link with any real life activity, used as teaching materials for students to work with. In teaching graphs and during most kinematics physics experiments in Ghanaian university contexts, students are mostly made to plot x-t, v-t or a-t graphs. Teachers usually extend their teaching by asking students to find the gradients of such graphs plotted and perhaps extrapolate or interpolate the graphs to find some other values. These teaching techniques of graphs have made it difficult for students when it comes to describing the shapes of kinematics graphs, conversion of kinematics graphs, getting the meaning and calculating slopes and areas under kinematics graphs, as they are not used to practicing these approaches in classrooms. They therefore misinterpret, especially in describing the shapes of graphs, find it difficult to change graphs from one form to another, give different interpretations to slopes and areas under graphs, and find it difficult to calculate the slopes and areas appropriately. In this study, how to use MBL tools and simulations to help students describe and transform kinematics (x, v, a-t) graphs, and how to use specific examples of kinematics graphs to get the meaning and calculate for slopes and areas under graphs, all in an interactive engagement teaching, will be considered. Results of the Ghanaian university students will be converted into mean proportion scores and compared to see how the use of MBL tools, simulations and graph samples in an interactive engagement teaching will improve on their knowledge in kinematics graphs.

2. Results from earlier studies

Graphical representations are important in students' physics education, especially during physics practical sessions, yet most students have limited understanding of graphs (Blume & Heckman 2000, Swafford & Brown 1989). Numerous studies have shown that students usually find it difficult to convey information with graphs and extract information from graphs (Swatton & Taylor 1994, Wainer 1992).

Students commonly misinterpret graphs as GAP (graph-as-picture), in which they expect the graph to be a picture of the phenomenon described (Beichner 1994, Bollen *et al.* 2016). Students interpret a graph of displacement versus time as if it were a road map, with the horizontal axis representing one direction of the motion rather than representing the passage of time. In problems dealing with balls rolling in tracks or people riding bicycles over hills, students using GAP will often draw velocity-time graphs resembling the shapes of the tracks or hills, rather than showing the velocity of the ball or bicycle (Murphy 1999). McDermott, Rosenquist and van Zee (1987) found that, even in the simple case of a straight line graph, the physics students in their study confused distance represented by the height of the graph with velocity represented by the slope of the graph in a position-time graph. The situation becomes more complicated when the graphs are curved, making the confusion of slope and height more common. According to Beichner (1994), other common difficulties students have when working with graphs are variable confusion, forming graphs from kinematics equations and graph transformation.

Though some students can calculate slopes, especially straight line graphs which start from the origin, others find it difficult when it does not pass through the origin of the graph. Most students cannot understand what the slope of a line graph connotes (Planinic *et al.* 2012). For example, students find it difficult to appreciate the fact that the slope of the velocity-time graph connotes acceleration of the object the graph is about. Also, students cannot tell which of two slopes is steeper (Beichner 1994, McDermott *et al.* 1987).

Similarly, a few students can calculate areas under kinematics graphs and could explain what these areas refer to. For instance, majority of students could not infer that the area under acceleration-time graph refers to the change in velocity of the particular object the graph is about (McDermott *et al.* 1987, Donnelly & Welford 1989, Eraslan 2008).

3. What do you intend to achieve?

The ability to comfortably work with graphs is a basic skill of the scientist. For example, graph construction, interpretation and transformation are very important and forms integral part of experimentation, which is the heart of science (Chambers *et al.* 1983, McKenzie & Padilla 1986). Graphs can provide a structured overview of the entire problem situation while still allowing details to be resolved. It has been found out that technology such as the use of microcomputer based laboratory (MBL) tools and simulations is promising for increasing students understanding and transformation of graphs (Mokros & Tinker 1987, Thornton & Sokoloff 1990). This will help students to engage in substantial reasoning to develop coherent understanding of graphs rather than fragmented ideas, which will not set students on the path towards learning. For example, allowing students to interact with motion sensors interfaced with computers to describe their motions could help students to make meaning of the description of their motion in relation to the graph plotted.

IOP Conf. Series: Journal of Physics: Conf. Series **1076** (2018) 012002 doi:10.1088/1742-6596/1076/1/012002

In University of Education, Winneba (UEW), the use of MBL tools in teaching is quite recent. The type of MBL tool used is called "Coach 6". It is a tool which is distributed by the CMA Science in Amsterdam for the active integration of computers in Science and Technology Education, with the view that this learning tool gives the science learner power to explore, measure and learn from the physical world (Kedzierska & Dorenbos 2007). It could also be used to plot graphs easily and quickly on the computer by physical movements of objects, transform kinematics graphs from one form into another (from position-time graphs to velocity-time graphs to acceleration-time graphs) and determine slopes of kinematics graphs.

In this study, we will investigate the effect of MBL tools, simulations and the use of already plotted graphs in the context of interactive engagement teaching to improve students' understanding of kinematics graphs in a group of first year physics students in a Ghanaian university.

The research question is:
"Could MBL tools, simulations and graphs samples used in interactive engagement teaching lead to students' conceptual understanding of kinematics graphs?"

4. What are the characteristics of an effective teaching approach in kinematics graphs?

After studying the literature the following sequence of activities were used in the teaching of kinematics graphs; *concept quiz, conceptual reasoning questions, interactive teaching, reflection, application and problem solving questions*. The purpose of using these activities has been summarized in Fig. 1:

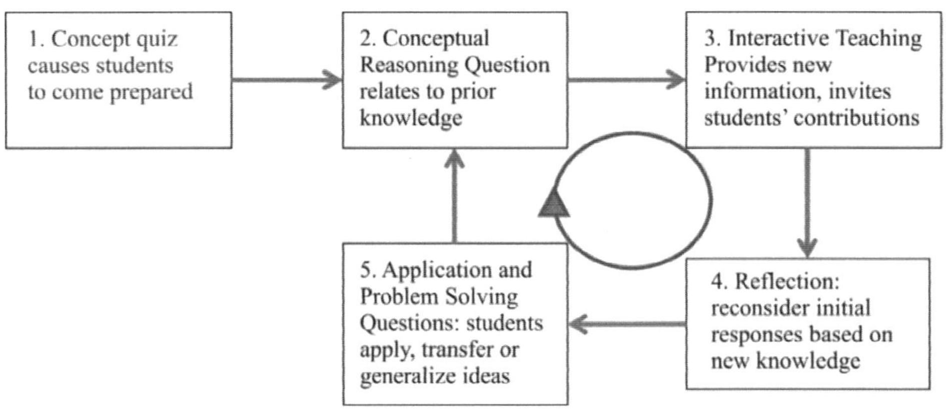

Fig. 1. Learning activities and purpose

5. Research setting

The research was carried out in the Department of Science Education, University of Education, Winneba (UEW). Participants in the study were first year physics students for two consecutive academic years of their first semester mechanics course. Thirty-seven (37) students were involved in the two academic years: 17 students in Group One and 20 students in Group Two. Understanding kinematics graphs is part of first year mechanics curriculum in UEW. Two out of the 11th-week lessons on mechanics were used to teach kinematics graphs. This consisted of six hours of teaching and four hours of problem solving session. The lecture room was equipped with computers, white board and a screen which allowed for the usage of beamers/projectors. The course was taught by one of the researchers. The research instrument used to gather data was the Test of Understanding Graphs in Kinematics (TUG-K) (Beichner 1994).

6. Method

Students were made to answer *(pre)* TUG-K a day before the teaching of graphs. During the day of lesson students were made to answer a concept quiz based on description of position-time graph, trans-

IOP Conf. Series: Journal of Physics: Conf. Series **1076** (2018) 012002 doi:10.1088/1742-6596/1076/1/012002

formation of position-time graph to velocity-time graph and calculation of gradients. The question is shown below.

Examine the movement of an ant running back and forth along a line in the graphs below.

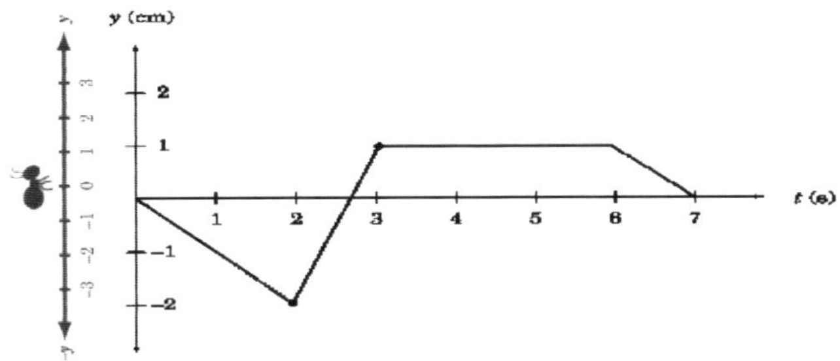

i. Give a brief interpretation of the position-time graph using the movement of the ant.
ii. Transform the position-time graph of the ant into a velocity-time graph.
iii. Determine the gradients of the various movements of the ant: t = 0-2 s, 2-3 s, 3-6 s and 6-7 s.

Transcription of how some students described the graph

1. **George:** Em!! The first part the, the ant's velocity was decreasing, and then it increased. The velocity then became constant and then decreased again.
2. [...]
3. **Peter:** The ant descended, climbed up till it got to a flat surface. It therefore moved for a while and descended again.

After a short discussion of the questions, students were made to answer a question on graph to activate their prior misconception on seeing a graph as picture (GAP). They were made to discuss in groups of four and choose the correct graph which best describes the question (*conceptual reasoning question*).

Little Johnny stands at the bottom of a small hill and kicks a ball. The ball rolls up the hill and then rolls back to Johnny. Which one of the following velocity-time graphs (A, B, C, or D) most accurately portrays the motion of the ball as it rolls up the hill and comes down?

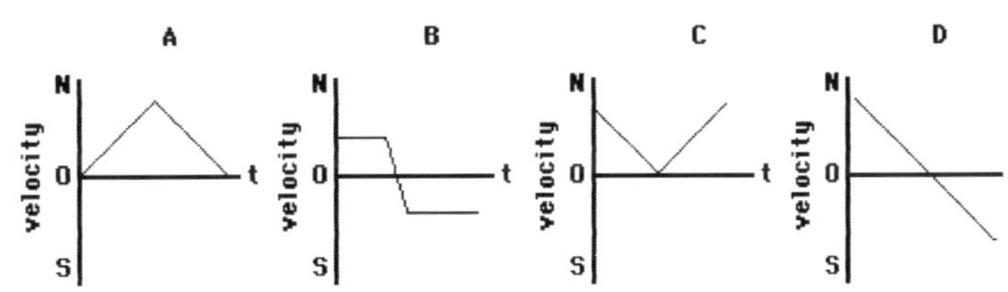

Transcription of students' explanations to the choice of their answer revealed that they see graphs as pictures (GAP)

4. **Class** (*responded*): "A".
5. **Teacher:** Anybody with different answer?
6. **Students:** (*responded*) no sir.
7. **Teacher:** You all chose "A'. Why "A"?
8. **Francis:** Sir, because the ball rolled northwards, then it returned back, which means that, it moves in opposite direction, it will be in southwards direction.
9. [...]
10. **Tony:** Sir A. Even the graph shows, because the 0 point, the velocity is 0. From the initial to a certain point or a certain height, northwards, then it came back...

The next activity was the interactive teaching, where students in groups of four were made to predict and practice with the MBL and motion sensors to plot graphs of their own movements (straight line graphs). This is shown below:

By the use of motion sensors/detectors and coach students are to observe displacement-time graphs, velocity-time graphs and acceleration-time graphs of the motions below by their movement;

- standing still; students analyze and describe the displacement-time graphs, transform it to velocity-time graphs and acceleration-time graphs of the motion by comparing the shapes of graphs.
- moving at constant speed in a specific direction; students analyze and describe the displacement-time graphs, transform it to velocity-time graphs and acceleration-time graphs of the motion by comparing the shapes of graphs.
- moving away and coming back at constant speed; students analyze and explain the shapes of the displacement-time graphs, transform it to velocity-time graphs and acceleration-time graphs of the motion by comparing the shapes of graphs.
- moving away and coming back with different speed; students analyze and explain the displacement-time graphs, transform it to velocity-time graphs and acceleration-time graphs of the motion by comparing the shapes of graphs.
- moving away, stopping and coming back; students analyze and explain the displacement-time graphs, transform it to velocity-time graphs and acceleration-time graphs of the motion by comparing the shapes of graphs.
- students were made to walk some already plotted x-t and v-t graphs.

By the use of simulations, students predicted and practiced curved x-t graphs and their transformations to v-t and a-t graphs. For example

- students were made to study different dot diagrams of x-t motions with a changing velocity (curved graphs) and their transformations to v-t and a-t graphs. Thus positive and negative changing velocities (slow to fast and fast to slow) were considered. http://www.physicsclassroom.com/
- They were also made to enter different values for initial position (m), initial velocity (m/s), acceleration (m/s^2) and time (s). Students were made to study the shape of the position time graph and transform the shapes to v-t and a-t graphs. http://www.physicsclassroom.com/

Students were made to calculate for the slopes and areas of already plotted graphs. Thus

- students were made to practice and determine the value of slopes of a straight line graphs in x-t and v-t graphs (graphs starting from origin and graphs not starting from origin). They were to determine what they were finding (*velocity, acceleration*).

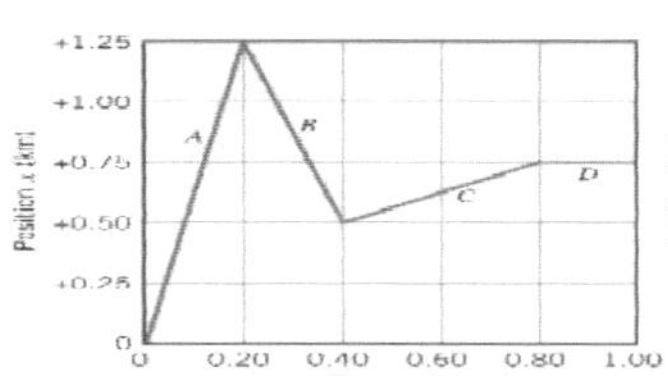
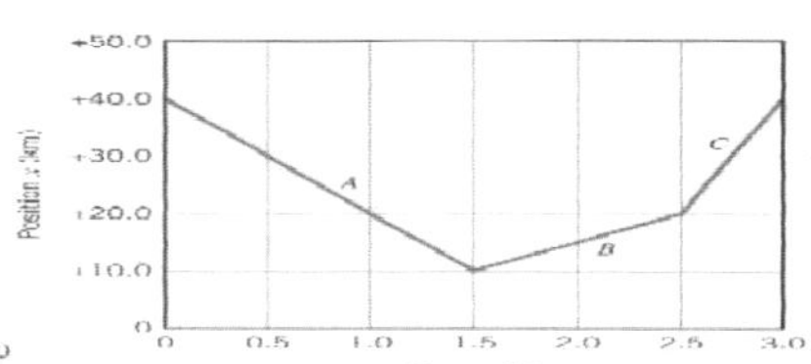
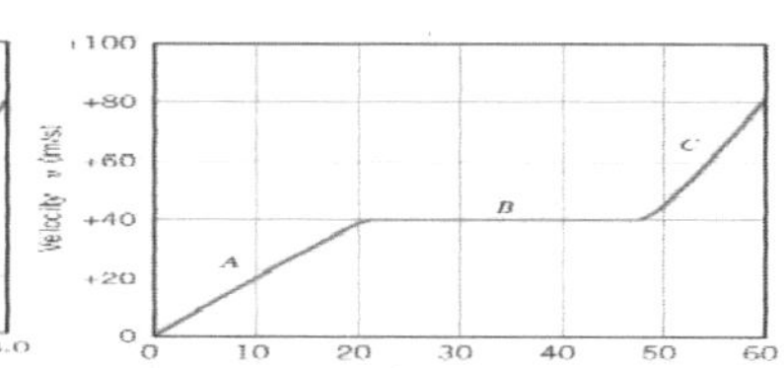

- students were made to practice and determine the areas under the straight line graphs of velocity-time graphs. They were to determine what they were finding (*total distance/displacement*).

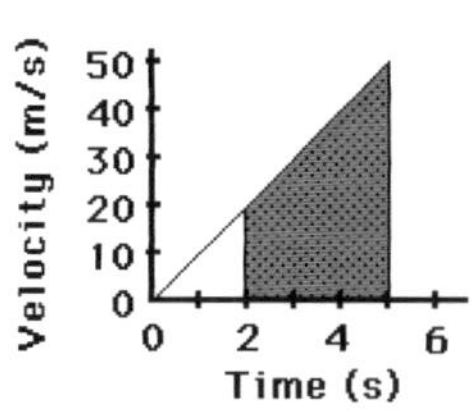

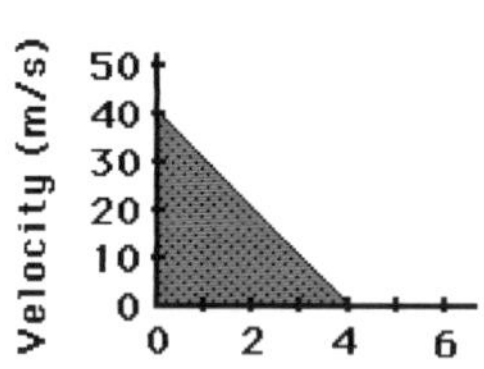

IOP Conf. Series: Journal of Physics: Conf. Series **1076** (2018) 012002 doi:10.1088/1742-6596/1076/1/012002

After allowing students to interact with MBL tools and simulations, they were made to reflect on their initial answers to the conceptual reasoning questions to see if they could improve on their answers (*Reflection*). All the students opted for "D" as the graph that accurately portrays the motion of the ball as it rolls up the hill and comes down. These were followed by application and problem solving where students could apply, transfer or generalize ideas relating to real world context. Some examples are given.

1. (a) Describe the following graphs.
   (b) Convert graphs 1, 4, 9, 10 & 11 into velocity-time and acceleration-time graphs.

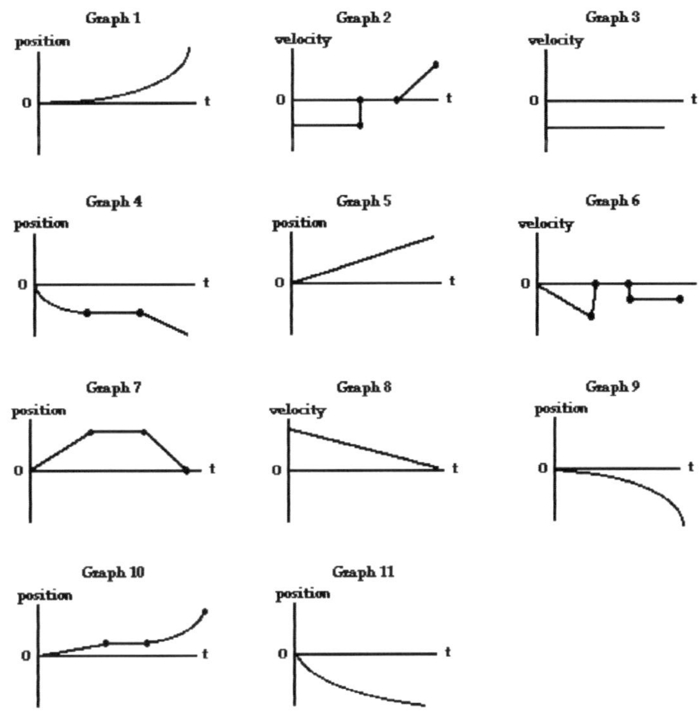

2. Give the meaning of the slope and area under the following graphs:

Graph	**Slope**	**Area under the graph**
position versus time	....	....
velocity versus time	....	....
acceleration versus time	....	....

3. The following graphs were plotted by the movement of the ant shown in the diagram. (i) From the shaded portions of the graphs determine the ant's displacement.
   (ii) Calculate the gradients of the 1st graph from t = 0-2 s, 2-3 s, 3-5 s and 5-7 s.

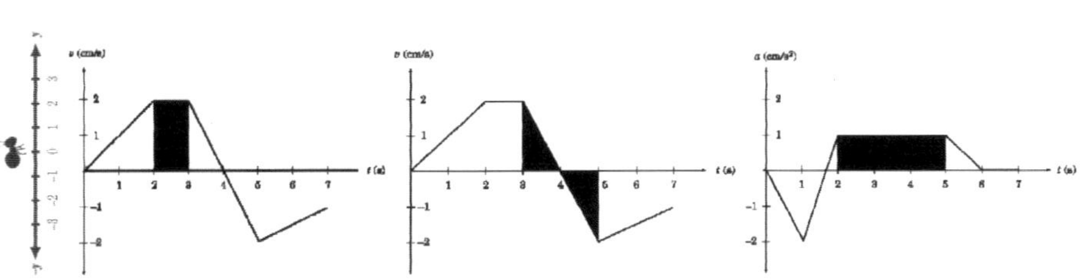

IOP Conf. Series: Journal of Physics: Conf. Series **1076** (2018) 012002 doi:10.1088/1742-6596/1076/1/012002

Students were made to answer *Post* TUG-K questions the day after the close of the lesson. The mean proportion scores of students' *Pre* TUG-K and *Post* TUG-K were used to calculate the Hake gain for each Group. The Hake gain values were compared for the Two Groups to determine their level of conceptual understanding in Kinematics graphs. This would help to see how the Ghanaian university students would be affected with the use of MBL tools, simulations and the use of already plotted graphs in the context of interactive engagement teaching in understanding of kinematics graphs. TUG-K is a multiple-choice standardized test, which consists of 21 questions with students common misconceptions as distracters. It is designed to assess students' kinematics graphing abilities. Students used 30 minutes in answering the questions in each session, due to some quantitative problems involved in calculating slopes and areas under graphs. TUG-K instrument was grouped into four main concepts under graphs and students mean proportion pre and post scores were calculated under these concepts: (i) Area under graph (ii) Slopes (iii) Graph description (iv) Graph transformations.

7. Results

To answer the research question, students were made to answer *pre* and *post* TUG-K questions. The mean proportion scores were used to calculate the Hake gain of students and presented in table 1. Students' mean proportion scores were calculated under the mentioned concepts and presented in tables 2, 3, 4 and 5.

Table 1. Mean proportion correct scores of *Pre* TUG-K, *Post* TUG-K, and Gain

Group	**N**	**TUG-K**		
		***Pre* (SD)**	***Post* (SD)**	**Hake Gain (SD)**
Group One	17	0.27(0.05)	0.69(0.07)	0.58(0.09)
Group Two	20	0.27(0.06)	0.71(0.07)	0.60(0.10)

Students' mean proportion scores in *pre* TUG-K was lower as compared with their post TUG-K. This was an indication that students did not have enough understanding of kinematics graphs before the beginning of the lesson. However, there was a considerable improvement in the mean proportion scores of *post* TUG-K. Also, the average normalized gain, $\langle g \rangle$, was about 0.6, which falls within Hake's medium-g courses, $0.7 > (\langle g \rangle) \geq 0.3$, which is a typical range for average effectiveness of courses in promoting conceptual understanding (Hake 1998). Hake developed his score for FCI but not for TUG-K, and along the same line we extended his way of calculation for TUG-K gain scores. The mean proportion scores of the two Groups were comparable.

Table 2. Comparing students' mean proportion correct scores in *Pre* and *Post* TUG-K- in area under graph (meaning and calculation)

Concept	**TUG-K Que. No.**	**Group**	**N**	***Pre* (SD)**	***Post* (SD)**
Area under the graph	1, 4, 10, 16, 18 & 20	Group One	17	0.28(0.02)	0.72(0.14)
(meaning & calculation)		Group Two	20	0.32(0.02)	0.74(0.08)

The performance of students in Pre TUG-K of both year groups was relatively lower as compared with that of their Post TUG-K. This indicates students' lack of understanding in areas under kinematics graphs at the beginning of the lesson. However, relatively higher mean scores of Post TUG-K show that concept of areas under kinematics graphs were highly appreciated by students after the interactive teaching.

Table 3. Comparing students' mean proportion correct scores in *Pre* and *Post* TUG-K- in slope (meaning and calculation)

Concept	TUG-K Que. No.	Group	N	*Pre* (SD)	*Post* (SD)
Slopes (meaning & cal.)	2, 5, 6, 7 & 17	Group One	17	0.46(0.03)	0.63(0.13)
		Group Two	20	0.49(0.02)	0.68(0.07)

Students' Pre TUG-K scores in both year groups were relatively lower; they were not as low as the other concepts. This is not surprising as most of the graph works that students do, especially in Ghana, are based on plotting graphs with given data, and calculating for the values of the slopes. Also, mathematics items in slopes are less difficult for students to solve (Planinic *et al.* 2012). However, their post scores were relatively higher, which is an indication that the intervention was helpful in enhancing students' understanding.

Table 4. Comparing students' mean proportion correct scores in *Pre* and *Post* TUG-K- in Graph description

Concept	TUG-K Que. No.	Group	N	*Pre* (SD)	*Post* (SD)
Graph description	3, 8, 9, 12, 19 & 21	Group One	17	0.22(0.02)	0.68(0.12)
		Group Two	20	0.20(0.02)	0.68(0.07)

Students' Pre TUG-K scores in description of graphs were relatively lower. This might be due to the way they saw graphs as pictures before the beginning of the lesson. There was a significant improvement in their scores in Post TUG-K after the lesson. This shows that the intervention in describing graphs might have had a positive effect on students' graph description (Thornton & Sokoloff 1990).

Table 5. Comparing students' mean proportion correct scores in *Pre* and *Post* TUG-K- in graph transformation

Concept	TUG-K Que. No.	Group	N	*Pre* (SD)	*Post* (SD)
Graph transformation	11, 12, 13, 14, 15 & 19	Group One	17	0.14(0.03)	0.70(0.05)
		Group Two	20	0.18(0.03)	0.70(0.05)

Students' Pre TUG-K mean proportion scores in graphs transformation were relatively lower. This is because graph transformation is rarely taught in Ghanaian schools. However, their relatively higher score in Post TUG-K might be attributed to the interventions in graph transformation during the lesson.

8. Conclusions

We have described how MBL tools, simulations and graph samples have systematically been structured into a sequence of activities in the context of interactive engagement teaching of kinematics graphs in the physics curriculum of a Ghanaian university. We noted more students' participation in class, high interactions among the students and also between the students and the computer. Students had high and almost equal gains in both year groups, which seem to confirm that the use of interactive engagement (IE) methods through MBL tools, simulations and graph samples in the classroom can increase graph teaching effectiveness well beyond that obtained in traditional practice (Hake 1998), though we cannot exactly know what really was most effective. Furthermore, students showed good conceptual gains in understanding of kinematics graphs, especially in getting the meaning and calculating for the area under graphs, knowing the meaning of slopes of graphs and also calculating for their values, description of graphs and graphs transformations. The difference of Post-TUG-K and Pre

IOP Conf. Series: Journal of Physics: Conf. Series **1076** (2018) 012002 doi:10.1088/1742-6596/1076/1/012002

TUG-K mean scores in both year groups was an indication that students' understanding in graphs have improved. Thus students had better understanding of kinematics graphs after the instructions in kinematics graphs with the use of MBL tools, simulations and graph samples.

References

Beichner R.J. (1994) Testing student interpretation of kinematics graphs. *American Journal of Physics* **62**, 750-762.

Blume G.W. & Heckman D.S. (2000) Algebra and functions. In *Results from the Seventh Mathematics Assessment of the National Assessment of Educational Progress* (Silver E.A. & Kenny P.A., eds.), Reston, VA: NCTM, pp. 269-300.

Bollen L., De Cock M., Zuza K., Guisasola J. & van Kampen P. (2016) Generalizing a categorization of students' interpretations of linear kinematics graphs. *Physical Review Physics Education Research* **12**, 1-10.

Chambers J.M., Cleveland W.S., Kleiner B. & Tukey P.A. (1983) *Graphical Methods for Data Analysis.* Belmont, CA: Wadsworth International Group.

Donnelly J.F. & Welford A.G. (1989) Assessing pupils' ability to generalize. *International Journal of Science Education* **11**(2), 161-171.

Eraslan A. (2008) The notion of reducing abstraction in quadratic functions. *International Journal of Mathematical Education in Science and Technology* **39** 1051-1060.

Hake R.R. (1998) Interactive engagements versus traditional methods: A six-thousand-student survey of mechanics test data for introductory physics course, submitted to *American Journal of Physics* **66**, 64-74.

Kedzierska E. & Dorenbos V. (2007) *Guide to Coach 6.* Foundation CMA/AMSTEL Institute, Universiteit van Amsterdam.

McDermott C.L., Rosenquist M.L. & van Zee E.H. (1987) Student difficulties in connecting graphs and physics: Examples from kinematics. *American Journal of Physics* **55**, 503-513.

McKenzie D.L. & Padilla M.J. (1986) The construction and validation of the Test of Graphing in Science (TOGS). *Journal of Research in Science Teaching* **23**, 571-579.

Mokros J.R. & Tinker R.F. (1987). The impact of microcomputer based labs on children's ability to interpret graphs. *Journal of Research in Science Teaching* **24**, 369-383.

Murphy L.D. (1999) *Graphing misinterpretations and microcomputer-based laboratory instruction, with emphasis on kinematics.* Retrieved from http://www.mste.uiuc.edu/Murphy/Papers/GraphinterpPaper.html.

Planinic M., Milin-Sipus Z., Katic H., Susac A. & Ivanjek L. (2012) Comparison of student understanding of line graph slope in physics and mathematics. *International Journal of Science and Mathematics Education* **10**(6), 1393-1414.

Swafford J.O. & Brown C.A. (1989) Variables and relations. In *Results from the fourth mathematics assessment of the National Assessment of Educational Progress* (Lindquist M.M. ed), Reston, VA: NCTM, pp. 55-63.

Swatton P. & Taylor R.M. (1994) Pupil performance in graphical tasks and its relationship to the ability to handle variables. *British Educational Research Journal* **20**, 227-243.

Thornton R.K. & Sokoloff D.R. (1990) Learning motion concepts using real time microcomputer-based laboratory tools. *American Journal Physics* **58**, 858-867.

Wainer H. (1992) Understanding graphs and tables. *Educational Researcher* **21**, 14-23.

IOP Conf. Series: Journal of Physics: Conf. Series **1076** (2018) 012003 doi:10.1088/1742-6596/1076/1/012003

Electricity and magnetism step by step and optics step by step: optional special labs in first years of pre-service teacher training

Irena Dvořáková, Leoš Dvořák

Department of Physics Education, Faculty of Mathematics and Physics, Charles University, Prague, Czech Republic

Abstract. Introductory physics courses for future physics teachers serve various, often quite "divergent" purposes. On one hand, they should present physics at university level, using mathematical tools like integrals, differential equations, tensors etc. with sufficient degree of exactness. On the other hand, they should develop conceptual understanding, physics insight and "intuition", skills necessary for doing experiments etc. Often, the development of such conceptual understanding and necessary skills is aimed also "at university level", assuming that students know the basics from secondary schools and have some experience with at least elementary experiments. Moreover, it is often assumed that students themselves will see the link between "secondary school physics" and "university physics". However, often this is not the case and future teachers complain that there is a large gap between physics at those two stages. To help students reduce this gap we created two optional seminars: *Electricity and magnetism step by step* and *Optics step by step*. These seminars give students better understanding of physical concepts as well as some basic methodical comments for their future work with pupils. Both seminars are based on the methodology of the Heureka Project (Dvorakova 2013a). In this article we would like to present our experience with both seminars.

1. Introduction

Our students – future physics teachers – have special physics lectures since the first year of their university studies. In the second semester there is a lecture (and seminar) on Electricity and magnetism. Students learn "the university physics" here, calculate different problems, etc. However, (about ten years ago) we found that students are able to successfully pass exams from this topic, but they have problems with very elementary tasks – to light up a bulb only with a battery, to connect parallel circuit etc. This was the main impulse which initiated a new seminar *Electricity and magnetism step by step*. The second seminar was opened two years later at students' request, because they considered a "step by step" seminar as very useful also in Optics.

2. Main goals of the seminars

Main goals the seminars should fulfil naturally follow from problems mentioned above. In general, students should there:

- build bridges between "university physics" and physics taught in junior and senior secondary schools in their minds and/or make such bridges stronger
- experience teaching and learning physics that is strongly based on simple experiments they do by themselves.

Of course, these two goals are interconnected: students should learn how simple experiments relate to relevant physics theory and should be able to illustrate such relations.

IOP Conf. Series: Journal of Physics: Conf. Series **1076** (2018) 012003 doi:10.1088/1742-6596/1076/1/012003

Moreover, seminars also give students a very good opportunity to become explicitly aware of some important misconceptions in the discussed topics by experiencing the misconceptions by themselves or by their colleagues.

Last but not least, each seminar is also a place where students improve their practical skills concerning simple experiments and, of course, some other more general skills and competencies: to collaborate, to discuss, to present results etc.

3. Basic characteristics of the seminars

As we already mentioned, both seminars are organized simultaneously with usual university lectures. They take 2 hours per week, in the 2nd and 3th semester. Although the seminars are voluntary, almost all students on adequate level attend them.

At the beginning of both seminars students are asked to solve several problems from basic school physics. They find that they have problems with this test; usually none of them are able to solve all tasks without mistakes. We know that students have many "empty places" in their minds, but because each of them has different "empty places", it has no sense to try only refilling missing knowledge. It is necessary to build concepts from beginning. That is why we start all topics from the basic level. Sometimes we say that no previous knowledge is expected (after the introductory test, of course).

Students work similarly to children at basic school. They do experiments, discuss, and solve different problems. Some examples of these problems will be presented later.

There are two more important parts of the program of seminars. One of them includes didactical elements and methodical comments for students' future teaching at school. In spite of the fact, that students have about four years of studying ahead of them and they usually have little idea about their future work with pupils, we consider those comments as very important. We are convinced that the sooner students gain a positive attitude to teaching the better. The second important part of the seminars is continuous interconnecting the basic level of physics with the university level.

The seminar *Electricity and magnetism step by step* includes four main topics – magnetism (permanent magnets and their field, magnetic field of the Earth), electrostatics (properties of the charge, electrical conductivity and electrostatic induction), electric circuits and electromagnetism (magnetic field of the wire and coil, mutual influence of a magnet and a conductor with current, electromagnetic induction). In the seminar *Optics step by step* we start with the basic properties of light, shadows and colours. Then students investigate mirrors (plane, concave and convex) and lenses. They study refraction of light, too. The last part of the seminar is focused on wave properties of light (diffraction, polarization, etc.).

Example 1 – Several tasks from the introductory tests

- Design and describe an experiment which shows that there are two types of electric charge. You can use a plastic rod, a container with a leaf of aluminum foil, insulation pad, and fur.
- Solve and give reasons for your decision: How does the ammeter reading change when the switch is closed?

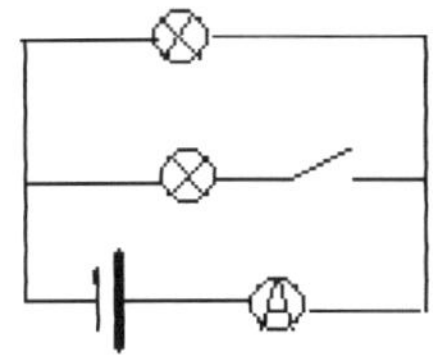

- Add two switches (plus wires if necessary) to the circuit so that their different combinations will cause 1, 2, 3, or 4 bulbs glow.

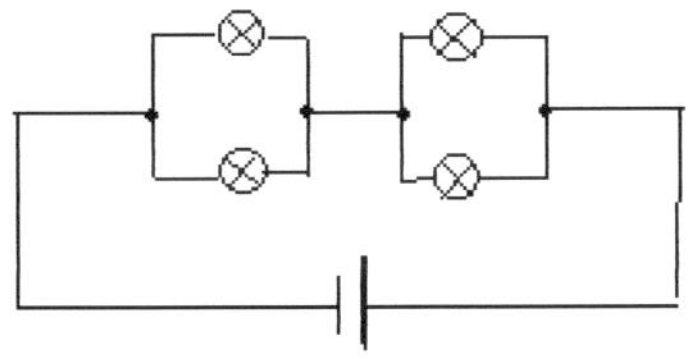

- Point A lies on an optical axis of a magnifying lens (not between the lens and its focus point). Find geometrically the image of the point A.
- Using magnifying lens we project a flame of a candle on the screen. Describe what you will see if we cover the top half of the lens.

Example 2 – Electric circuits

The results from different research studies show that students need more opportunities to work with real electric circuits. Students have problems with understanding and correctly applying the concept of a complete circuit. They also need to understand multiple representations of the circuit to deeply understand its behaviour, (e.g. Osborn 1983, McDermott & Shaffer 1992). Therefore we spent several hours going through the whole methodological sequence, which helps students to build the concept of electric circuits using multiple representations. This sequence was described in details in Dvorakova (2013b).

In this text we present only the starting part of the sequence:

Step	Activity	
1.	Play with a small bulb and a battery, connect bulb to a battery (use a piece of wire if necessary) and try to make it glow (Exp. 1).	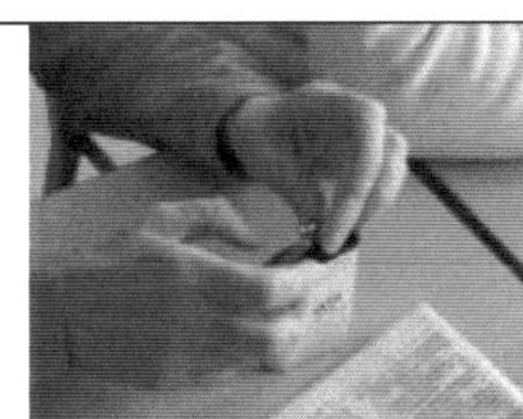
2.	Draw a picture which describes the arrangement of your experiment.	
3.	Connect the bulb to the battery through as many things as possible at the same time so that it glows (work in pairs) (Exp. 2).	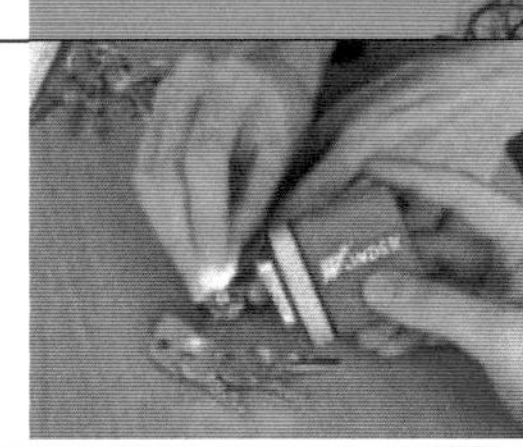
4.	Sketch how the experiment looked (using 4 – 5 pieces is enough). Describe the common properties of things you used in the Exp. 2.	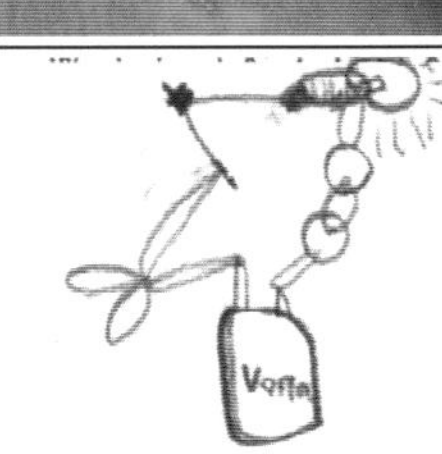

<table>
<tr><td>5.</td><td>Draw different situations, where the bulb is connected to the battery, yet it does not shine (crazy ideas).</td><td></td></tr>
<tr><td>6.</td><td>Teacher shows circuit diagrams to represent circuits.
Teacher shows tables for describing the state of switches and bulbs.</td><td><table><tr><td>S</td><td>B1</td><td>B2</td></tr><tr><td>0</td><td></td><td></td></tr><tr><td>1</td><td></td><td></td></tr></table>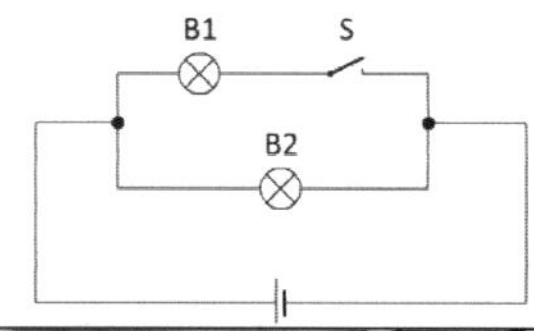
</td></tr>
<tr><td>7.</td><td>Students work with real bulbs, switches and batteries and build assigned circuits.</td><td>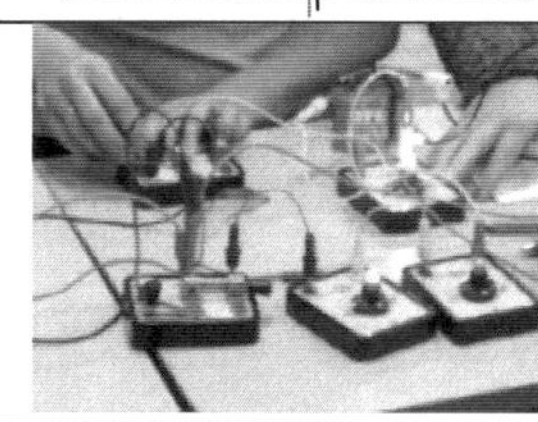</td></tr>
</table>

Fig. 1. First activities from the methodological sequence Electric circuits

We would like to bring your attention to step 3. This competition proved to be a very interesting activity and students like it very much. Furthermore it gives a lot of possibilities for further conclusions.

Example 3 – outdoor activities
One of the very favourite activities is outdoor playing with mirrors and lenses. We need a day when the sun in willing to play with us.

Students do a set of experiments, for example:

- Reflect the sunlight to the given place using one plane mirror.
- Reflect the sunlight to the given place using two plane mirrors.
- Light up a piece of paper using a concave mirror.
- Light up a piece of paper using a magnifying lens.

Fig. 2. Outdoor activity during the seminar *Optics step by step*

4. More than mosaic of experiments

Examples presented above could possibly raise an impression that the core of the seminars is just a "mosaic" of simple experiments, perhaps loosely connected with parts of theory. However, they are much more than that. In fact, looking from a broader view, they serve also as classes preparing students for their future teaching career.

Structure of the seminars follows the way how these topics are taught at junior secondary level by one of authors – so students are, partly "implicitly" and partly openly, being prepared for teaching of these topics. This "horizontal structure" of the seminars is supplemented by "vertical connections" to formulas, quantities and theory students learn in university lectures. (As it was mentioned, the lectures are taught in the same semester as the "step by step" seminars.)

The very important aspect of the seminars concerns the fact that learning students experience there is inquiry-based. (It is not completely free inquiry, the role of a teacher is important there but it is definitely far from any cookbook approach.) Of course, the inquiry involves discussion of students and students groups, their collaboration, presenting results to colleagues etc. Therefore, later in their teaching carriers, students can use methods they experienced there as "templates" in their own teaching.

5. Feedback from students

Though no formal tests or quantitative surveys were done at the end of seminars, various forms of feedback from students show that the seminars fulfil the needs expressed by goals mentioned above.

First, in spite of the fact that the seminars are optional, they are attended by nearly all students (future physics teachers). Students of older classes recommend the seminars to their younger colleagues as really useful. At more formal level, such attitude is proved by a survey our Faculty does at the end of each semester among students. (In last few semesters the results of the survey are used to choose and acknowledge "best educators" of the Faculty – and the ratings of our seminars reached that level.)

Apart from quantitative evaluation some opinions of students they wrote in the survey can illustrate the impact of seminars. Let's present here just a few examples of student's views:

> *I think that this seminar prepares us very well for our future teaching career. I appreciate that we solve all problems at the level of students of junior secondary school, to be able to explain them physics in their own words.*
>
> *The style of teaching was very inspiring. ... It forced us to think.*
>
> *I surely recommend it to clarify one's knowledge. We did a lot of experiments and tried to explain them.*
>
> *Very good seminar from optics, it really makes sense. It is good to connect optics from lectures with physical "playing" with lenses, prisms etc., seminars are supplemented by thinking about interesting phenomena and instruments. ... anyway, the seminar is an excellent choice.*

Apart from these views expressed several weeks after the end of each seminar there is also a long-lasting effect. In classes aimed at physics teaching which take place about two years after the seminars, students repeatedly use what they learned in the seminars and explicitly refer to them.

Based on all this feedback, though mostly qualitative one, we think we can state the seminars really fulfil the needs that forced us to create them.

6. Conclusions

It is hopefully not necessary to describe detailed structure of the seminars here. (To really go into details would make this text much longer.) We think that every educator can adapt the content to their own needs. It is the main approach which is important.

In case you think about creating similar seminars or you have been organizing them already and you are willing to share experience, you can contact us for further details.

References

Dvorakova I. (2013a) Active learning in the Heureka Project – teachers in the role of students. *Proceedings of the Conference ICPE-EPEC 2013*, 46-63.

Dvorakova I. (2013b) Electric circuits in The Heureka Project – multiple representations. *Proceedings of the World Conference on Physics Education 2012*, Pegem Akademi, 239-247.

McDermott L.C. & Shaffer P.S. (1992) Research as a guide for curriculum development: An example from introductory electricity. Part I: Investigation of student understanding. *American Journal of Physics* **60**(11), 994-1003.

Osborne R. (1983) Towards modifying children's ideas about electric current. *Research in Teaching and Technological Education* **1**, 73-82.

IOP Conf. Series: Journal of Physics: Conf. Series **1076** (2018) 012004 doi:10.1088/1742-6596/1076/1/012004

Best practices for a good laboratory experience

Daniela Marocchi, Marina Serio, Marta Rinaudo

Physics Department, University of Turin, Italy

Abstract. After the investigation on students' perception towards the laboratory activities, presented at GIREP-MPTL International Conference 2014 (Marocchi & Serio 2015), we conducted a new analysis concerning the aspects of support and enhancement of the teaching activity in laboratory.
We investigate i) how students prepare the laboratory activities, ii) the importance of the presence of teacher, technicians, tutors throughout the entire laboratory process, iii) the usefulness and ease of use of the informatics instrumentation.
This second phase involves first year students during the academic year 2015/16. Results of questionnaires highlight the importance of teaching methods used, as well as of all the professional figures involved during the educational laboratory experience.

1. Introduction

The perception that students have towards laboratory activities has been presented in a previous work (Marocchi & Serio 2015), based on results of a questionnaire for students of different ages and school levels. In that case, we paid attention to the development of laboratory interest and capabilities starting from high school up to the third university year of study. We investigated several aspects, such as comprehension of the physics concepts, interest in laboratory activities, complementary nature of laboratory activities and of classroom lectures. However, within the open comments of questionnaires, other aspects seem to need further attention: for example the didactic material and the laboratory data sheets, the preparation of students that help as tutors (in our case the tutors are university students who help in the acquisition of data) and the presence of technicians. In particular, we want to analyse the importance of various professional figures present in laboratory and the utility of the educational path proposed to the students: didactic and computer materials, on-line homework, auto evaluation tests, etc...

In this paper, we examine the formative impact of student-tutors and of technicians, which in laboratory are complementary to teachers. Students, although they sometimes regret the possibility of managing autonomously the practical part, are well aware of the necessity of a guide. Nevertheless, in order to achieve maximum understanding from the students, it is crucial to know how tutors and technicians work in relationship with them. We also study the use and utility of the assessment tools in ongoing and final evaluation. Other analysed questions are about the possibility to use the instruments and the usefulness of computer equipment.

We wrote a questionnaire for students of the first year of Physics during the academic year 2015-2016 at Turin University (Italy). The survey was limited to the 150 first year students in order to assess also the impact from the different teaching methodologies used in high school. We also proposed a questionnaire to the other persons that are present in laboratory during the course, i.e.: technicians and tutors. We present and discuss here the results, in order to highlight how teaching methods as well as all the persons involved in the experience of educational laboratory are important.

IOP Conf. Series: Journal of Physics: Conf. Series **1076** (2018) 012004 doi:10.1088/1742-6596/1076/1/012004

2. Students' opinion

The laboratory course proposed to our first year students lasts overall six months, with two periods each lasting ten weeks and a central pause of four weeks. The contents are a theoretical part on 'statistics and data analysis techniques and a laboratory part with twelve laboratory experiences'. The purpose of the experimental part is double: it gives both the possibility to apply statistical methods to real data (instead of doing theoretical exercises on not-real data) and to verify some important laws presented in the parallel Physics course. The laboratory experiences proposed in the first and in the second module differ in the complexity of the analysis needed to reach the results. In fact, in the first module, the objectives of the experience are often the outcome of direct measurements; in the second module, the results derive from many direct measurements assembled. In both cases, it is required that the students have the capability to apply the techniques of data analysis presented during the course. A second important objective of the laboratory activities is to increase the ability of working in-groups, to organize the work, and to reflect on the obtained results.

We administered a questionnaire at the end of each learning period, articulated according to the characteristics of each module. The laboratory course of the first year at Turin University is very demanding: 12 ECTS out of 60, which is the total number of ECTS required during the completely academic year. The difficulty is even higher given that many students have never had any experience of laboratory activities during high school (see also Marocchi & Serio 2015).

Within the studied sample, 33% of students had never attended a laboratory activity before enrolling in the university and that 13% had done some lab work but had not needed to complete lab reports on that work; 60% had never used a spreadsheet for analysing and graphically showing the results. Only 19% of the students said that they often wrote laboratory reports in high school. For the other students (80%) the principal reason was that they had never gone to the laboratory or that they had seen only qualitative experiences carried out by the teacher.

The relationship with all the people involved in the laboratory activity results are very important. More of half (60%) of the students appreciate the availability of people like tutors and teachers, while technicians remain marginal in their experience (Fig. 1).

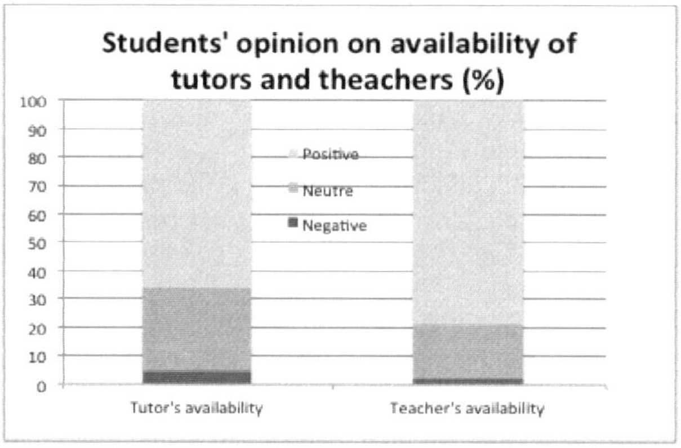

Fig. 1. Students' opinion on availability of tutors and teachers

In the questionnaire, part of the questions concerned the general aspects of laboratory activity such as:

- the development of practical activities (such as the ability to correctly use scientific instruments, to properly measure and estimate the error to be associated, to graphically report the results, to critically review the results of the statistical analysis);
- the type of experience (physical laws to verify through data analysis);
- the weight of the course, in terms of time and personal student work;
- the appropriateness of lesson's contents for the performance of the experiences.

A. Practical skills development
B. Please Type of experience
C. Autonomy in managing the experiment
D. Duration of experience and analysis sessions
E. Time to devote to drafting reports, compared to the number of ECTS
F. Appropriateness of lesson contents for the performance of experience

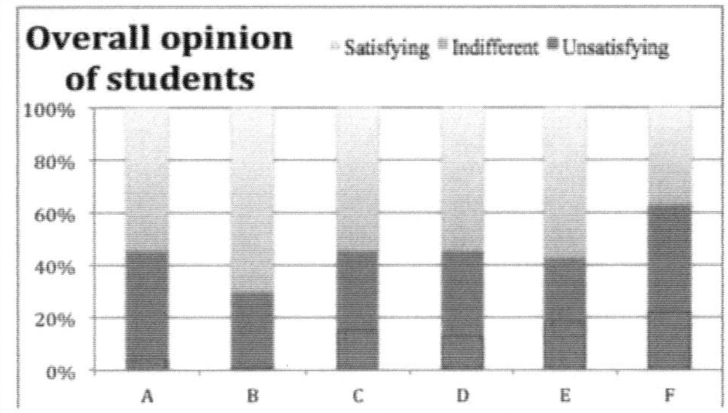

Fig. 2. Student's opinion related to general aspects of laboratory activities

Some aspects (Fig. 2) are judged partially inadequate (from 15% to 22%): in particular the time spent for writing the reports, the little autonomy in managing the experiments and the appropriateness of the explanation provided during the lessons for the implementation of the laboratory experience. Our considerations as regard these critical points are:

- It is the first experience in report writing for many students, so they perceive the report preparation to be hard and laborious, both in the first and in the second writing after the teacher correction. In fact, it often requires revisions concerning not only numerical data but also linguistic expression in the scientific field.
- As for the autonomy, taking also into account the poor experience of the students and the complexity of the used instrumentations, the proposed experiences require a tutor for the experimental part and the presence of the teacher for the robust analysis phase.
- Appropriateness of the explanations provided during the lessons is a very delicate point. The theoretical presentation of the experiences takes place before the start of the laboratory sessions. Since it is impossible to move the instrumentation into the classroom or to be in the laboratory with the students, it is very difficult to provide operational details. Moreover, due to the number of students, shifts have to be established and a part of the students performs the laboratory experience even a few weeks after the explanation.
- A self-assessment questionnaire was prepared for each experience with 5-6 multiple-choice questions and immediate feedback. In the questionnaire are non-present open questions because they require a longer time for compilation. Questions concern the goals of the experience and some of the operating procedures presented in the lessons. The student must answer the questionnaire before going to the lab. Students also evaluate this self-assessment activity: 80% of them (Fig. 3) consider the self-assessment questionnaires and the feedback useful to help the review of the lessons. The questionnaires are now under review to improve clarity.

A. are useful for recalling the crucial points presented at lesson
B. need too long to answer
C. are not clearly expressed
D. feedback is useful to understand errors
E. more open questions would be helpful

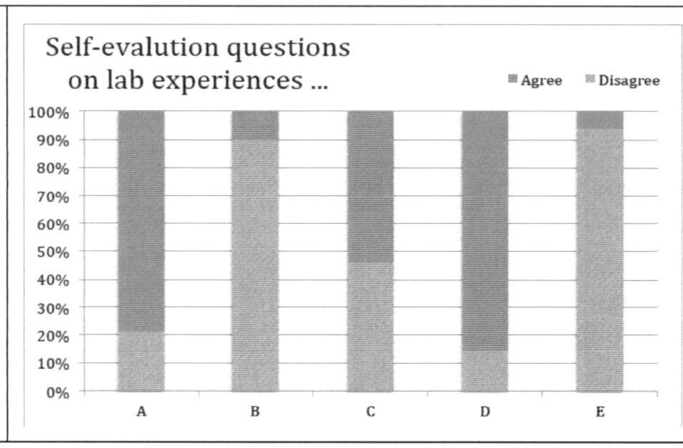

Fig. 3. Student's opinion on self-evaluation questionnaires regarding laboratory experiences

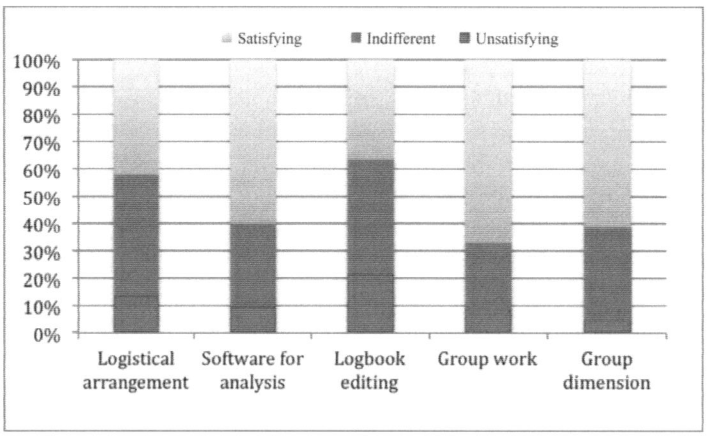

Fig. 4. Student's opinion relative organizational aspects of laboratory activities

Finally, we investigated some organizational aspects of the laboratory (Fig. 4): the logistic arrangement, the group-working mode, the logbook editing and the software for the analysis. The two aspects considered inadequate by more than 10% of the sample are logistics and logbook editing. Indeed, during the 2015-2016 courses, the Department considerably rearranged the spaces reserved for the laboratory, with a real discomfort for both students and teachers. As far as the logbook is concerned, many students are not accustomed to report in a concise but complete way what happens during the experience. They often consider only important to record the numeric data directly on Excel spreadsheet or Mathematica notebook to make analysis with the computer. Hence, they usually forget to note details that may be useful during the analysis phase, the discussion of results and the critical conclusion of the work.

3. Tutors' opinion

Tutors are bachelor or master students in Physics who receive a scholarship to assist students in the laboratory practical tasks, since teachers cannot follow all the students at the same time.

During the academic year 2015-16, half of the student-tutors had performed this task for the first time.

Before the start of the course, they attended an educational training on technical aspects of the laboratory experiences, which is their main task. This preparation period is brief but balanced. Moreover, technicians are available to help with every technical problem encountered during laboratory sessions. Furthermore, students know that, as for problems in the analysis of measured data, they have to refer to the teachers. Tutors' opinion on the effectiveness of this initial training is not uniform: half of the student-tutors think to be not able to explain the importance of some measure procedures (50%), or to clarify the in-depth analysis (88%) that students are requested to do, or to have the correct didactic approach.

The work of the student-tutor does not only represent a source of help for teachers and technical staff. There is a common understanding (71%) that it can also be an important formative training for those students who become tutors. So many student-tutors ask a specific formation in software and in didactic procedures. They express that, by being tutor, they have a deeper comprehension of physics topics and that they have the opportunity to became leaders of a working group. Some students that work as tutors perceive as important also a specific training on didactic aspects because they are interested in understanding better the work of a teacher. These answers reveal the usefulness of this experience for their future working choices.

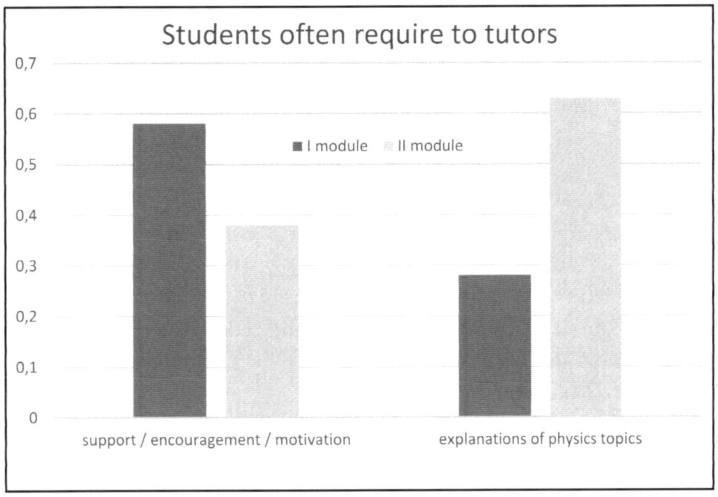

Fig. 5. The figure shows the different type of request from students to tutors during first and second part of laboratory course

With regard to the differences observed during the two course modules (Fig. 5), tutors reveal that during the first part of the laboratory sessions they often have to encourage students at their first laboratory experience (58%). Instead in the second part of the course the more important role is a guide during the technical operations (63%) because of the greater complexity of the experimental task.

Tutors positively note the presence and efficiency of technical staff during the laboratory sessions (71%), and the availability to give further indications (83%). Tutors have noticed the revised laboratory-sheets (100%), the clarity of instructions (83%), and the availability of the teachers to explain in depth the practical tasks relative to the laboratory activities (67%). They have (50%) an uncertain opinion on the teacher's availability to give educational didactic training; therefore, this point needs to be improved.

4. Technicians' opinion

Persons who have PhD in Physics and participate to research groups in Physics Department compose the technical staff for this laboratory course at Turin University. Therefore, they are very special persons. Their role includes i) the correct preparation of instruments and ii) the technical formation of tutors (Fig. 6).

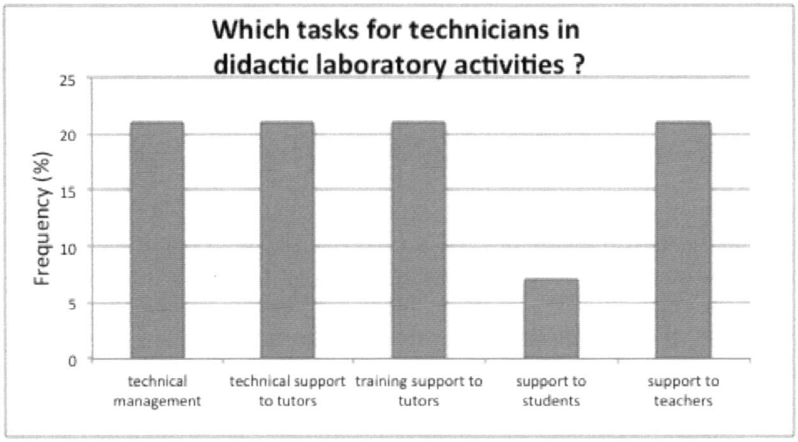

Fig. 6. The figure shows what technicians think about their laboratory task

Even if technicians say that the didactic laboratory is a very interesting task (100%), they also declare that it is not satisfying their expectations and capabilities. Moreover, they (60%) desire more knowledge in educational subjects and a greater didactic collaboration with teachers (60%) in order to be familiar with the educational objectives of various laboratory experiences. Furthermore, technicians (80%) propose to have a wider possibility of interaction with students, not only with tutors.

5. Materials

In order to focus on the materials that we used as support for the teaching activity in laboratory, we inserted in the questionnaire for students several questions on laboratory materials and on procedures

Students think (80%) that the didactic material (available on e-learning platform Moodle) has good quality and that it is complete (Fig. 7).

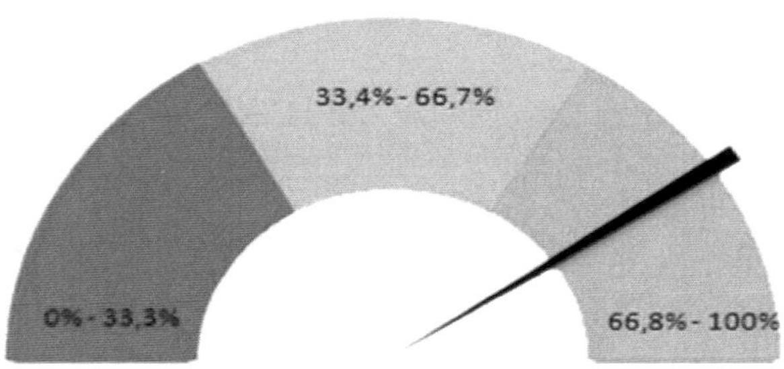

Fig. 7. Students' opinion on didactic material is essentially good

One element of paramount importance for a positive laboratory activity is the preparation of students before the laboratory sessions. Moreover, it is also an important factor for the success of the experience. Some instruments used in didactic environment are self-evaluation tests, online exercises, open questions, etc. In particular, students have the possibility, before the laboratory session, to read the monograph, which reports the physics of the experience and some technical procedural notes. This material is available on-line. Among the students, 90% use the monograph and 82% answer to self-evaluation tests, 72% reads lesson notes and 64% asks for information from other students that have previously done the same experience.

In order to test the competences acquired, we have prepared multiple-choice on-line tests for every experiment: students had to answer to these questions before the laboratory session. Feedback helps (87%) students to evaluate their preparation: the time needed to answer is not too long (87.5%), so

IOP Conf. Series: Journal of Physics: Conf. Series **1076** (2018) 012004 doi:10.1088/1742-6596/1076/1/012004

they can easily do it after the class lesson and before the laboratory session. About 57% of the sample judges that some questions are not sufficiently clear. Hence, teachers will have to make an effort in revising them. The mark of these tests does not enter in the final valuation, but we note that they are an important incentive to increase the attention of students, which feel interested in going more prepared to laboratory. The tutors (83%) who noticed an improvement in students' competence after the use of multiple-choice tests have confirmed this impression also.

We thought the auto-evaluation tests to prepare students before the laboratory work and the questions essentially concern the method with which they have to operate. Students considered them useful also for the preparation of the final exam (87%), but some students say that they would prefer technical questions (27%) or questions of physics (48%).

We note a correlation between a good and regular execution of the multiple-choice tests and examination result. Table1 shows that all the student of the course that had the maximum results (30 cum laude in Italian University) did well all the auto evaluation tests. At the same time, none of the students with bad results in the auto evaluation tests has reached the maximum examination result; Fig. 8 shows the results for student of the B course versus the number of auto evaluation tests completed.

Table 1. Number of valuation 30 cum laude during the exam (maximum evaluation in the Italian University) versus number of well-done auto-evaluation tests (course A)

number of well-done tests	number of valuation: 30 cum laude
0	0
1 to 3	0
4 to 5	0
all	6

In addition, students considered the correction and the return of the first three laboratory reports before the final report useful for a good preparation (96%), but part of the students (25%) encounters difficulty to complete the reports during the didactic period.

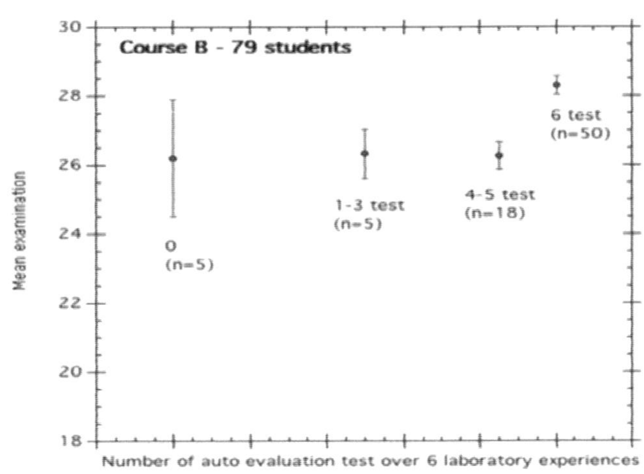

Fig. 8. Results of the examination versus the number of auto-evaluation tests done (course B)

6. Conclusion

We have examined, through a questionnaire offered to 150 students, student-tutors and technicians, the importance of all the professional people involved during didactic laboratory sessions. The results of the questionnaire have stimulated a reflection on formative activity, so we have introduced some good practices in the course during the present academic year: we present them briefly, considering them useful for teachers in their work in class.

In the course, students appreciated the type of physics experiences proposed, the development of practical skills, their increase of informatics capabilities by using PC for the analysis of the data, group work and availability of tutors and teachers. For 33% of them this course is the first laboratory activity and 60% of them had never used a spreadsheet for analysis nor have used graphics to show some results. For this reason, students have many difficulties in writing the reports. To support them in this task, we have prepared for the first experiment an online format in which student can insert data, results, comments, and in-depth analysis. We organized the format in sections that correspond to the different items of a scientific report, so it can help as a guide for the writing. Therefore, we will provide the students with some old reports and the revision form used by teachers for correction and evaluation. In this way, students can see what the teacher looks for in the text and in the analysis. They can use the form to mark the old report and then as a guide to correctly compose their own.

The importance of a logbook and of its correct compilation will be highlight during lessons and tutors have to control its proper use. During the pause between the first and the second module, the teacher marks the logbook and gives it back. In the future, we think it will be useful to consider tools such as Google Drive or Google class, to share online files and generate an e-logbook easier to manage. The high number of students attending the course is currently the greatest limitation of this solution.

The work of the student-tutor not only acts as a help for teacher and for technical staff, so the use of student-tutor has to be encouraged. There is a common understanding (71%) that it can also be an important formative training for those students who become tutors. So many student-tutors ask a specific formation in software and in didactic procedures. To improve the tutor training, we decide that each tutor becomes an expert on one of proposed laboratory experiences. The training ends with a presentation of the experience made in front of the teacher before the beginning of laboratory sessions. During this presentation, the teacher also discusses with the tutor the didactic aspects of the experiment.

All the technicians say that the didactic laboratory is a very interesting task; furthermore, they want a greater didactic collaboration with the teachers (60%) in order to be familiar with the educational objectives of various laboratory experiences. Therefore, we engage also the technicians, interested to deepen the didactic content of the experiments, in the final step of tutors' training.

Students substantially appreciate (80%) the laboratory team and the quality of materials offered on Moodle platform. We note that auto-evaluation tests are a good instrument for student's preparation before the experiments, but students considered them useful also for the preparation of the final exam (87%) and we note a positive correlation between a good and regular execution of the multiple-choice tests and examination results. A statistical analysis has shown which pre-experience questions are less clear. We have set up a review of the questions and subjected the new versions to a small sample to verify the clearness of the text, with particular attention to the formulation of the incorrect answers. During the laboratory session, the teacher shows and discusses with the student wrong answers in order to strengthen the effect of automatic feedback. Just at the start of the first period teacher informs the students that every partial evaluation is important for the final one. Therefore, we want stimulate a regular study. We are now developing an appropriate evaluation form to take into account all the partial results correctly.

The relationship with high school teachers is very important to support them in the hard work of moving their student close to the physics laboratory activities. The starting situation revealed by the questionnaire push us to share our formative thoughts with high school teachers, to encourage them to

IOP Conf. Series: Journal of Physics: Conf. Series **1076** (2018) 012004 doi:10.1088/1742-6596/1076/1/012004

a greater use of laboratory teaching, and in November 2017, we realized a meeting where we have discussed with them the problems encountered during laboratory activity.

References

Marocchi D. & Serio M. (2015) Laboratory activities and the perception of students. *Proceedings of GIREP-MPTL International Conference 2014*, 1059-1070.

IOP Conf. Series: Journal of Physics: Conf. Series **1076** (2018) 012005 doi:10.1088/1742-6596/1076/1/012005

Effectivity evaluation of experiments in physics education by memory retention

Karel Havlíček

Charles University in Prague

Abstract. Evaluating effectivity of various forms of experiments and practical work in physics lessons is a very complex problem. The goals that we as educators would like to achieve by practical work are broad. Examining all the aspects of practical work and evaluating them would be very difficult to do from a technical standpoint. It would be very time consuming, challenging to implement in normal school courses and in turn very costly. The aim of this study was a development of relatively simple method that would give insight into what methodologies of practical work are effective in teaching high school students' conceptual knowledge. Based upon the framework for considering the effectiveness of a practical task proposed by Abrahams and Millar (2008), we believe that this information can be ascertained by examining which experiments students remember and to what extent. We expect that more effective methodologies of performing classroom experiments will leave more permanent and complex imprints in the student's mind and therefore these experiments should not only be remembered more often, but also to a greater detail. To this end a relatively short questionnaire was developed, consulted with experts and piloted, that focuses on finding out what types of experiments do students remember from the last six months of their physics education course. The Questionnaire is constructed in such a way, that it doesn't specifically work with any part of physics curriculum and therefore can be used across all school institutions and all school years without any modifications. Validity of the data can and should be increased by cross-referencing gathered data with information gained by interviewing the teachers of the respondents. Gathered data should also allow us to map what general types of students exist in regard to their relation to practical work and if certain methodologies are more effective when used on different groups of students. Hopefully, these findings will give us some insight into what forms of practical work are actually effective and if so, then on what type of students.

1. Introduction

As stated above, the evaluation of effectivity of practical work as a teaching tool is a difficult issue. This is due in no small part to the high expectations we as educators might have when introducing it in the classroom. The goals we would like to achieve by practical work are broad, as stated for example by Hodson (1990) and Bennett (2004), and range from mere motivation to teaching good science practices. Research of practical work effectivity in teaching all of these goals would be too complex, time consuming and in turn too expensive to do. For that reason I decided to focus on only one of these goals and that is teaching conceptual knowledge. Furthermore, there was no possibility for me to conduct a controlled experiment that would put sufficient number of students through a physics course constructed in such a way that it would contain selected experiments taught in ways we would also select while maintaining a control group. To find enough willing teachers to partake in an experiment that would require from them so much additional work without any monetary gain is nigh impossible. With these limitations in mind, I opted to conduct research in such a way, that it would not require me to force the teachers to use preselected experiments and methodologies.

IOP Conf. Series: Journal of Physics: Conf. Series **1076** (2018) 012005 doi:10.1088/1742-6596/1076/1/012005

2. Theoretical framework

As a basis for my work I chose the framework for considering the effectiveness of a practical task proposed by Abrahams and Millar (2008). In it he describes his model of the design process and evaluation of a practical task as a four stage endeavour. My research focuses on the transition between the third and fourth stage, e.g. between "what the students actually do as they undertake the task" and "what the students learn as a consequence of undertaking the task" (Abrahams & Millar 2008). However, how can you evaluate what students learn, when you can't control what topics are taught? This lack of curriculum control makes creation of tests for evaluating particular concepts very difficult?

This issue had to be tackled from a different angle. Based upon the first levels of Bloom's taxonomy of cognitive domains, students should be able to recall and apply knowledge they learned. These two tasks can therefore form a viable probe into students learning. However, question arises if evaluation of memory retention can in turn help evaluate effectivity of an experiment. Our method is built upon the current knowledge of inner working of human memory and memory loss of school knowledge. As stated by Semb and Ellis (1994):

> *Any theory about loss of information learned in school should be more concerned about loss of semantic content than episodic content.*

The same authors also write:

> *As new information is assimilated, existing knowledge structures/schemata should be modified and extended. For example, when prior knowledge is repeated during instruction, overlearning should occur. This should result in higher levels of retention.* (Semb & Ellis 1994)

Since academic knowledge is ideally incorporated into a scaffolding of previously learned knowledge and interacts with it to form a coherent web of information, we expect an effective experiment to be linked with knowledge taught in such neural scaffolding. Therefore, such experiment should help in recollection of the knowledge taught and vice versa. Complexity and quality of such mental construct should positively affect the memory retention, therefore an experiment that teaches concepts effectively should also be more memorable.

However, we acknowledge that even ineffective experiments could be very memorable due to explosions, social aspects and other novelty factors. So we conclude that pure memory retention of an experiment isn't a sufficient sign of its effectivity as a teaching tool. Recollection of the experiment has to be accompanied with the recollection of concepts involved and possibly with the ability to use those concepts.

3. Research questionnaire

In order to get enough teachers to volunteer themselves and their pupils for our research, its method had to be implementable with as little effort for them as possible. Therefore, a questionnaire was selected as a primary data gathering device. Measured data would be triangulated and cross-checked by interviewing the teachers afterwards. Additionally, the questionnaire for students had to be relatively brief, or we would risk that significant number of students wouldn't answer it seriously or even answer it at all. For those reasons a two page limit was set with a 20-minute time frame in mind. The questionnaire was consulted with experts and went through two separate pilot runs. In its current final state it consists of two distinct parts.

Low-level retention

The first part and also the first page of the questionnaire maps the low level retention of experiments. Students are asked to compile a list of experiments they remember from last six months of their current physics classes and put each of them in a specific column depending on what methodology was used. Students are presented with following classes.

Table 1. Classes of experiments as used in the questionnaire

A	B	C	D
Experiments performed by teacher	Showing video recorded experiment (YouTube, DVD, ...)	Laboratory work – Significant teacher supervision	Laboratory work – you manage your own work

This list is then evaluated in two ways. First and foremost, how many experiments from each class does this student remember? And additionally, since the name of the experiment is constructed by the student, how many of those names contain an idea or concept as opposed to how many were based upon an observable world. For example the same experiment was called "A truck and a ball" by one student and "Conservation of momentum – track, ball" by another. The usage of a concept in the name can be symptomatic of the student forming a connection between that particular experiment and the scientific concept. This link between the domain of ideas and the domain of observables is also part of Millar's framework.

High-level retention

The second part of the questionnaire deals with in-depth recollection, testing of conceptual knowledge and its application. Students are asked to choose one experiment from their list which they believe they remember the best. Chosen experiments are then searched for any similarities like:

- Chosen experiments are\aren't recent
- Were repeated the most
- Were of particular class (A, B, C, D as above)
- Were named using a concept

Students are also asked to fulfil following tasks:

- List all the equipment that was used for that experiment
- Describe their observations from that experiment
- Name the concept shown by that experiment if there is any
- Explain the shown concept

Gained information is then evaluated for its correctness and completeness. Since these are open ended questions, their evaluation is open to interpretation. To increase reliability of the results, all filled questionnaires are independently coded by two physics teachers (i.e. experts) in accordance with a written codebook. Both evaluations are then crosschecked and any discrepancies are discussed until both coders are in agreement.

4. Preliminary results

The questionnaire was deemed satisfactory after the second pilot run and therefore the data from this pilot phase were incorporated into the data gathered in the "live" inquiry. While data were gathered from about 200 students, they are in various phases of the coding process and so far none of them have been fully analyzed. All the presented preliminary data (from around 30 subjects) are from the second pilot run. Since this data come from a single class taught by one physics teacher, it should be noted, that these results may not represent general school population. Because of that I opted to showcase only a few select pieces of data, data I find interesting, instead of presenting the whole analysis. That I shall publish at a later date when the statistical sample is more significant.

First data I would like to present is an overview of the distribution of experiments listed in the first part of the questionnaire among their classes. Data are shown in Fig. 1.

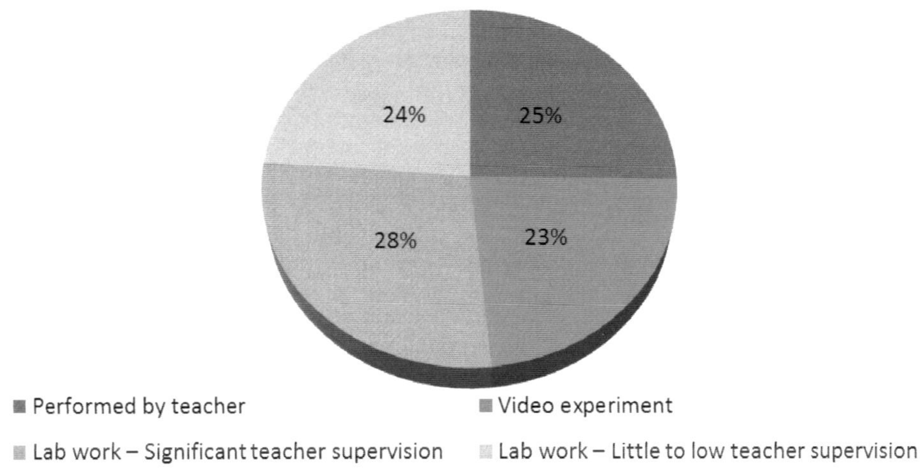

Fig. 1. Overview of the distribution of experiments listed in the first part of the questionnaire among their classes

As you can see, there isn't any significantly preferred class. However, two of the classes represent ways of performing a laboratory work in class. When joined together, this newly created general class occupies 51% of remembered experiments. This conforms with broadly accepted theory that lab work is more memorable and possibly more effective as a teaching method. When prompted to select the best remembered experiment, students generally chose experiments from the same classes as seen in Fig. 2. Interesting is also a complete avoidance of video experiments in the second part of the questionnaire. This leads me to believe that they were either very ineffective in this particular class or difficult to describe and/or explain and in turn scarcely chosen.

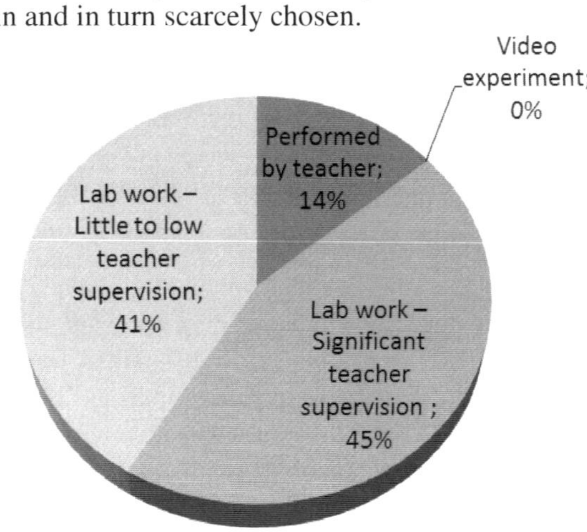

Fig. 2. Distribution of experiments chosen as best remembered

Final data we present is the retention period of chosen experiments. Which in this particular setup means how long it is since the students saw the chosen experiment. Given that memory retention slowly decreases over time, it would be logical to assume that without any other effects the best remembered experiment should be the most recent one. However, data shown in Fig. 3 show that this is rarely the case. Students preferred to describe experiments that were 1-6 months old. Actually, the most frequent time period was 3 months.

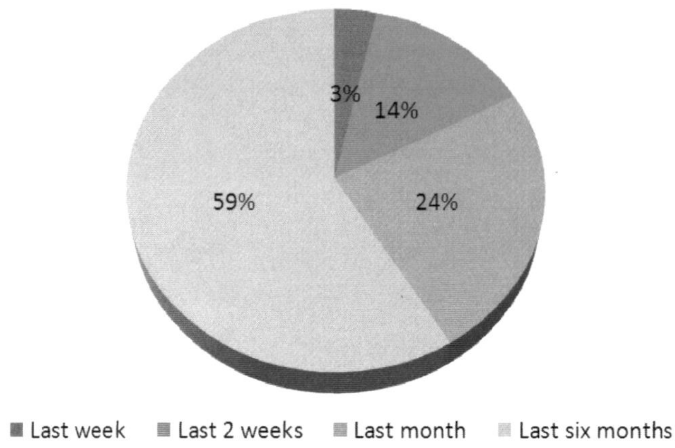

Fig. 3. Time since last exposure to an experiment that student selected as best remembered

This leads me to believe that when done correctly, practical work can, to a certain degree, overcome natural memory degradation.

5. Future plans

Data gathered from the live inquiry will be fully coded, triangulated for added reliability and analyzed. We hope to map both the student types present in classroom in regard to the practical work and the effectivity of various elements of practical work. After the initial mapping is complete, I hope to follow it with additional study that would selectively introduce different teaching methodologies into the lectures. The same questionnaire would be then administered to ascertain the effectivity of those methods.

Acknowledgement

The author would like to thank doc. RNDr. Leoš Dvořák, CSc., RNDr. Martina Kekule, Ph.D and RNDr. Mgr. Vojtěch Žák, Ph.D. for their guidance and expertise when designing the questionnaire. This study was supported by the Charles University in Prague, project GA UK No 1164216.

References

Abrahams I. & Millar R. (2008) Does Practical Work Really Work? A study of the effectiveness of practical work as a teaching and learning method in school science. *International Journal of Science Education* **30**(14), 1945-1969. https://doi.org/10.1080/09500690701749305.

Bennett J. (2003) Teaching and learning science. A&C Black.

Semb G.B. & Ellis J.A (1994) Knowledge Taught in School: What is Remembered? *Review of Educational Research*, ProQuest Central **64**(2), 253-286.

Hodson D. (1990) A critical look at practical work in school science. *School Science Review* **70**(256), 33-40.

Content transformation for experimental teaching nanoscale science and engineering to primary teachers

Leonidas Manou[1], Anna Spyrtou[2], Euripides Hatzikraniotis[3], Petros Kariotoglou[4]

[1]University of Western Macedonia, Department of Primary Education, Florina, Greece
[2]University of Western Macedonia, Department of Primary Education, Florina, Greece
[3]Aristotle University of Thessaloniki, Physics Department, Greece
[4]University of Western Macedonia, Department of Early Childhood, Florina, Greece

Abstract. This study is a response to the need to examine the educational perspective of the Nanoscale Science and Engineering field (NSE) in all grades of education. Specifically, we address to educate primary teachers about essential concepts, applications and phenomena of the NSE discipline. In order to do so, we first seek to establish, the salient concepts, principles and phenomena that should be introduced to primary teachers, as a result of the transformation of the content, to content appropriate for the target population. In addition, to address the challenges of making the formatted content structure accessible to primary teachers, a variety of experimental activities were designed, which we present in detail.

1. Introduction

Nanoscale Science and Engineering (NSE) is a modern field of research. It deals with the ability to manipulate matter at the nanoscale level, and exploit the unique properties and phenomena that occur at this scale in order new, revolutionary applications to emerge. An unprecedented rate of advancements is taking place ranging from biomedical applications, electronics and advanced materials (Jones *et al.* 2013).

The explosion of the NSE achievements requires a commensurate response by the education community. Trying to confront with this challenge, initiations regarding NSE education have been made across the world in all grades of education (Feather & Aznar 2011).

However, the inclusion of NSE to education seems to be a challenging task. Several issues emerge, such as the determination of the content to be taught, teacher training and the design of the experimental activities that would make the NSE content accessible to learners (Hingant & Albe 2010, Jones *et al.* 2013).

2. Theoretical background

Features of NSE discipline

There is no denying that NSE is an interdisciplinary field that focuses on the study and the manipulation of matter at the nanoscale, a realm below 100 nanometers approximately (Hingant & Albe 2010, Logothetidis 2012). The development of advanced techniques and sophisticated instrumentation, such as Electron Microscopy and Atomic Force Microscopes (AFM), has paved the way for the emergence of the NSE field, as they provided scientists and engineers with new capabilities for the imaging, manipulation and characterization of nanoscale materials (Murty *et al.* 2013, Jones *et al.* 2013). A large number of applications has emerged ranging from electronics, advanced materials, medicine and so on (Hingant & Albe 2010).

The true foundation of NSE lies on the alteration of materials' properties at this dimension (Murty *et al.* 2013). Indeed, most definitions articulating about NSE emphasize the remarkable properties that matter exhibit at the nanoscale (Ramsden & Freeman 2009, Hingant & Albe 2010). When the size of a bulk material is gradually reduced, its properties remain the same at first. Then some small alterations may occur, until, the material's size reaches the nanoscale regime, where dramatic changes in properties can happen. For example, a macroscale chunk of gold is a familiar yellow metal. When this piece of metal is broken into smaller centimeter – long pieces, it will still appear yellow. However, if the pieces are broken down about million times, the tiny gold nanoparticles will vary in color, ranging from blue (50 nm in size) to orange (1 nm in size), depending on their dimension (Murty *et al.* 2013).

Why NSE in education?

Several types of argumentation justify the inclusion of NSE in education. The first one involves the shortage of appropriated skilled workforce in nano-related sectors It has been estimated, that there will be a worldwide demand for 6 million workers in NSE fields by 2020 (Feather & Aznar 2011). Considering that elementary school students may constitute the future workforce of NSE, an as soon as possible interaction with this emerging field should be a primary focus (Lin *et al.* 2015). Moreover, teaching about careers in science is argued that it can offer new opportunities for the young students to expand the range of their aspirations (Osborne 2008).

The concern of ensuring a trained workforce may grow due to the documented decline of student's interest towards STEM studies and related careers as well. This drawback may be attributed either to the fact that a looming gap exists between the science that students are learning in schools and their everyday life or because teaching students about modern scientific and technological advances is often underestimated (Kähkönen Laherto & Lindell 2011). On the other hand, it is argued that teaching students about NSE advancements can be a motivator to excite them in the classical fields of physics, chemistry and biology and direct them to pursue studies and careers in STEM (Healy 2009, Jones *et al.* 2013).

The second type of argumentation stems from the explanation of structure – property relation, which is a common theme to all grades of education. Traditional school curricula focus on macroscopic properties first and lead to explanations on the atomic level. However, many of the macroscale phenomena we experience in our everyday life demand explanations on the nanoscale level. For example the superhydrophobic property of some plants has its origin to the hierarchical nanoscale structures (Bhushan 2016). To conclude, not only the sub- nanoworld, which is characterized by atoms and molecules, has an impact to our lives, but also the nanoworld, can determine the properties of the macroscale materials that we manipulate.

In addition, a third type of argumentation is relevant to the scientific – technological literacy, or nanoliteracy (Jones *et al.* 2013). Nanoscale materials have already been penetrated to products that we utilize in our everyday life. For example, nanoscale metal oxide nanoparticles are already used in diverse applications such as sunscreens and antimicrobial coatings (Ramsden 2009). However, it is questionable whether nanoparticles can be a menace for the human body when they digested or released to the environment (Bhushan 2016). An informed citizenry is needed to appreciate risks and put them into context.

This study deals with the above issues. Especially, we focus on training Primary Teachers (PTs) about this modern field. A major consideration is to establish the content structure for instruction and build upon related activities. In order to do so, we first sought to map the core concepts of NSE discipline across all levels of education, and with content transformation process to form the content to be taught to PTs. We aim to answer to the following question:

Which of the fundamental NSE concepts can form a content structure for educating PTs and what laboratory activities can be built upon to develop NSE content learning?

3. Methodology

As a first step, we sought to detect through the literature, the fundamental ideas that construct the NSE discipline, across all levels of education, from elementary school up to undergraduate level. We followed this route because it is noted that teachers need to develop understanding about the fundamental concepts in order to learn NSE (Jones *et al.* 2013). Especially, we searched for studies that were conducted with the definite aim to determine the NSE concepts that should be introduced to a particular grade level of education.

Major data sources were books and journals, which we collected with several ways. We examined web pages of journals whose area of focus was educational, e.g Journal of NanoEducation. The key words written to the search engines of these journals included terms as nanoscience, nanotechnology, primary/secondary/post-secondary, nanoscale, fundamental/essential/core concepts, engineering nanoeducation and combinations, e.g. k-12 nanoeducation. By writing the above terms, we searched for relevant material into electronic data sources such as scienceDirect, ERIC (The Education Resources Information Center), GoogleScholar, and google. Furthermore, we examined the references of the studies collected, gathering more articles, until no new study was found.

Blonder & Sakhnini (2016) and Sakhnini & Blonder (2015) published results concerning a similar research. However, the distinction of our research is that we do not seek for differences and similarities among concepts that emerged from separate studies concerning the same grade level (e.g concepts within the secondary level), but only among different grade levels. For this reason, we integrated the concepts that came from different investigations but were addressed to the same educational level. In addition, we weren't interested in comparing the items that were included under the umbrella of each concept. Instead, the included items served as a tool to recognize the underlying essential concept. For example, "size and scale" was explicitly identified as a core concept for secondary and tertiary education. While this concept is not explicitly mentioned as such in the primary education, the included items of the core concept "nanoscale definitions" (Huang, Hsu & Chen 2011) led us to "correspond" the latter concept with the concept "size & scale" (table 1).

We traced four studies that satisfied the criterion we have established: one for the academic level (Wansom *et al.*, 2009), two for the secondary level (Stevens *et al.* 2009, Blonder & Sakhnni 2016) and one for the primary education (Huang *et al.* 2011).

The second step after identifying the basic concepts of the NSE discipline is to form the content structure to be introduced to the PTs. In order to do so, we had to take into consideration not only solely the above identified NSE concepts, but also the students' and teachers' perspectives about the particular content as well (interests, difficulties in understanding, preexisting knowledge, etc.) (Duit *et al.* 2012).

4. Results

In the following, we present the fundamental NSE concepts that may form a content structure as well as the related intended aims for educating PTs (table 1). In parallel, we focus on the laboratory activities that can support the introduction of the NSE content.

Concept 1: Size and Scale (Lesson 1 & 2)

Size and Scale is considered as fundamental for the comprehension of all of the NSE concepts (Delgado *et al.* 2015). Size refers to the actual extent or amount of something, and nanoscale can be defined by the size of the objects that NSE focuses on. As size and scale change, not only the properties of matter change, but the models, the tools, or the dominant forces that underlie these changes may alter (Stevens *et al.* 2009). To illustrate the significance of the particular concept, Jones *et al.* (2013) stress: "If teachers lack a fundamental knowledge of the size and scale of nanometers, it is not clear how they can understand and teach students about how materials behave differently and how tools and techniques differ when working at this small scale" (p. 11).

Recommendations regarding the conceptualization of "size and scale" suggest five levels of thinking (Magana *et al.* 2012): a) categorizing, involves grouping together objects that have similar sizes into categories b) ordering, involves putting several objects in order of size, for example from the biggest to the smallest c) qualitatively relating objects, which can be described by the statement "the difference in sizes between object A and object B, is approximately the same as the difference in sizes between object C and D" d) quantitatively relating objects, which is described by the statement "the difference in sizes between object A and object B, is about so much in number as the difference in sizes between object C and D" and e) expressing the numerical absolute size of an object with standard measurement units, e.g. the size of bacteria is 1 micrometer . The first three levels refer to size and the latter two to scale cognition and all of them can be contextualized in NSE learning goals.

Concept 2: Tools (Lessons 1-5)

Research has revealed that learners meet challenges regarding the above conceptualizations (Delgado *et al.* 2015). For example, undergraduate students classified several objects in two categories according to their visibility (Magana *et al.* 2012). This may hinder their understanding about the objects that NSE focuses on. In order to confront with this challenge, we decided to introduce the concept of "*tools*".

The different tools that scientists utilize when they study objects of several scales, can be implemented as a criterion for grouping several objects into the scales. Although the idea "tools" has been characterized as inadequate to teach students about the properties of materials at the nanoscale (Jones *et al.* 2015), we decided to include it to the to be taught content, as it can be implemented as a "stepping stone" for the discrimination of the three scales (e.g. macroscale-naked eye, microscale – optical microscopes, nanoscale – electron microscopes) (Stevens *et al.* 2009). This treatment of the concept "tools" led us to neglect introducing to PTs how the tools at different scales operate (especially at the nanoscale), or the different kind of non – optical microscopies (e. g AFMs, SEMs) that are utilized for studying the nanoscale objects.

Regarding lesson 1 & 2 we designed activities in order to train teachers considering size & scale – tools. Teachers had to complete the task that can be seen to image 1. The image shows a diagram, which is called Scale ladder. The black arrows point the absolute sizes of specific landmark objects, the white circles the tools that render these objects visible, and the three large frames (with three different colors) indicate the three scales: the macroscale, the microscale, the nanoscale. PTs watched two videos that introduced the nanoscale via macroscale and microscale. For example, one video presented the absolute size of a human being, then of a bacterium and then of a DNA, and the tools that are used for observation. Teachers were assigned to write next to the appropriate arrow, the object, and into each circle, the tools that are utilized. Then they had to name the colored boxes as macroscale, microscale, nanoscale. PTs were encouraged to refer to and complete the scale ladder with several objects that would encounter during the following lessons (e.g nanobumps – lesson 6-7, nanopores – lesson 9) in order to enrich the macroscale, microscale or nanoscale areas.

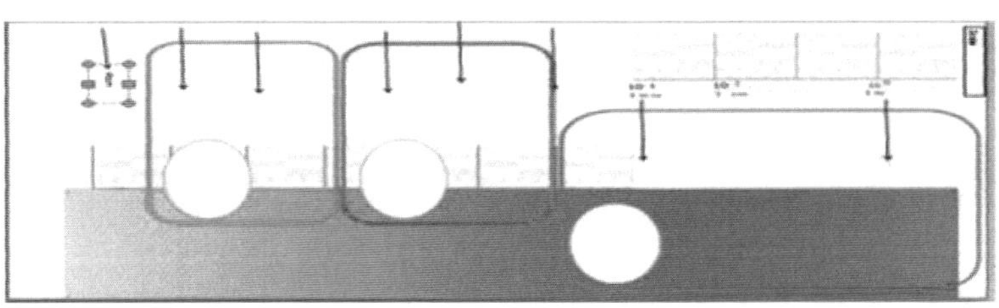

Fig. 1. The Scale ladder, which was used as a tool in order PTs to develop understanding about size and scale – tools

Concept 3: Model (Lessons 3-9)

One of the main reasons, that learners encounter difficulties for developing understanding about NSE, is the lack of intuition (Xie & Pallant 2011). A higher level of thinking is acquired, which will not only be implemented just for knowing facts, such as how small a nanometer is or what the structures of some nanomaterials look like, but, even more importantly how materials behave at the nanoscale or how nanoscale systems are engineered. The nanoscale world is foreign to the learners as it is the world that is governed by electromagnetic forces and quantum mechanics, and the phenomena that occur at that scale are often counterintuitive (Xie & Pallant 2011, Jones *et al.* 2013). In order to make nanoscale accessible to Pts, we turned our attention to the concept "*models*" (Stevens *et al.* 2009, Wansom *et al.* 2009). Indeed, it is acknowledged by scientists, engineers and educators, that models and modeling can be a valuable tool for learners in order to develop understanding of the phenomena and processes that take place at the nanoscale (Schonborn *et al.* 2016).

However, models are approximations of reality and as such, they can provoke misconceptions about the targets they represent (Turkoglu & Oztekin 2016). For example, it has been noted that often when students think of microscopic objects, they recall pictures from textbooks they have seen. In this case, students may develop macroscopic conceptualizations of microscopic objects (Tretter *et al.* 2006). Such a way of thinking may prevent learners to develop a firm grasp of the concept "size and scale".

For this reason, we considered that aspects about the nature and role models should be introduced to PTs. In particular, we were based on the overview by Oh and Oh (2011) who identified important aspects about models such as the meaning of a model, purposes of modeling, multiplicity of scientific models, change in scientific models and uses of models in the science classroom. In addition, we prompted teachers to build models to represent nanoscale entities and phenomena, as it is documented that the process of creating models can improve the understanding about the nanoscale phenomena (Daly & Bryan 2010). For example, teachers constructed their own models to represent the structure of nanoscale objects (virus or DNA) & outline the viral infection (image 2).

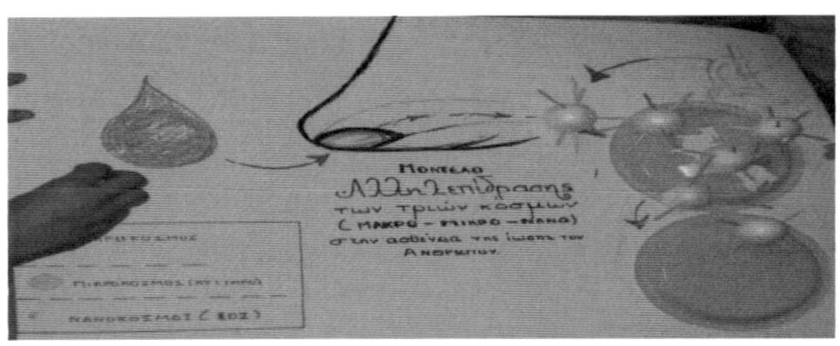

Fig. 2. A model of viral infection which was built by one group of teachers.
The viruses are represented with ball and stick models and the cells by the yellow plastic dishes

In general, the content of models was woven in parallel with the NSE content (table 1). To be specific, aspects about the nature of model were introduced to PTs gradually as they mentally made the transition among scales.

Concept 4: Size dependent properties (Lessons 6-8)
Across literature, it is stressed that much of the advances of the NSE field have become a reality, because of the unique phenomena and the remarkable properties that emerge at the nanoscale. It is argued that the statement "a material's properties may change with size" consists the essence and foundation of the NSE discipline (Hochella 2002). At the same time, most of the articulated definitions of NSE emphasize the unique properties that materials exhibit at the nanoscale (Ramsden & Freeman 2009).

Accordingly, it seemed incumbent to introduce the concept of "size dependent properties". A material's properties can be size dependent either because quantum effects begin to dominate in the nanoscale regime (e.g. quantum confinement), or because of the dramatic increase of the ratio surface area to volume (S/V ratio). Research concerning the negotiation of this concept, has revealed several issues. In particular, it seems that learners encounter difficulties when they attempt to explain why a material's properties change at the nanoscale. For example, secondary science teachers are inadequate to explain the size dependency of the quantum dots color (Bryan *et al.* 2012). In addition, it is acknowledged not only by NSE educators but even from secondary teachers, that implementing quantum physics to explain the occurrence of nanoscale phenomena is a challenging task and may be inappropriate even for the secondary level (Laherto 2011). Taking into account that "quantum mechanics is an extremely complex subject that requires extensive experience in both mathematics and science, and its counterintuitive predictions are difficult to grasp even for expert scientists" (Stevens *et al.* 2009, p. 33), we chose not to negotiate at all, to the PTs, properties that emerge due to quantum effects.

Continuing the above discussion, we placed our attention to properties that can be explained in terms S/V. However, it is reported in the literature that students' from elementary up to undergraduate level and teachers' applications of surface area to volume relationships in several contexts is problematic. For example elementary school teachers have trouble understanding the S/V beyond its mathematical calculation (Swarat *et al.* 2009). Similar remarks have been reported in other ratios. For example, in the case of density which is defined as the mass per unit volume, the difficulties, that student encounter, are mostly qualitative and conceptual rather than quantitative (Spyrtou *et al.* 2008). Bryan *et al.* (2012) assumed that secondary students may not have yet developed the mental skills to apply S/V relationships in NSE contexts.

To confront the above challenges, we introduced the S/V ratio qualitatively avoiding using the relevant mathematical ratio. We took into consideration that the amount of S/V may define the contact area of two surfaces. That is, for the same volume, the smaller the size of a material the more increased exposed surface area will be created, which in turn will increase the amount of contact between two surfaces. For example (image 3), we present in the context of the lotus effect, the decreased surface contact area between a spherical water droplet and the surface of the lotus leaf. Furthermore, PTs quantitatively measured in situ on printed images depicting plant leaves, the contact angle between a water droplet and a surface using a protractor (image 4). In addition, via gecko effect, we demonstrated the impact of the increased surface contact area to adhesion (lesson 8).

Fig. 3. A video simulation that shows the small surface contact area between a water droplet and the nanoscale structures of the lotus leaf

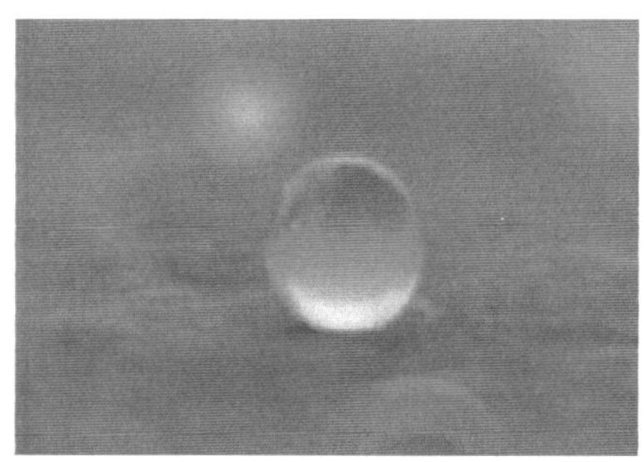

Fig. 4. One of the printed images that was used in order PTs to draw and measure the contact angle by a protractor

Concept 5: Applications (Lessons 5-9)

Sakhnini & Blonder (2016) argue that NSE applications can form a platform for the inclusion of several core concepts, such "as size dependent properties". Introducing NSE applications to school classrooms that are meaningful to student's everyday life may motivate them and increase their interest towards NSE learning (Lin *et al.* 2015). This approach to introduce NSE concepts via applications has been proposed by several researchers in the nanoscale field education (e.g. Lin *et al.* 2015 primary education, Delgado *et al.* 2015 secondary education).

Based on the above consideration, we wove applications with the concepts of "size and scale" and "size dependent properties". For example, after PTs established knowledge about size and scale, they contextualized this knowledge within a viral infection case, in which a nanoscale object (virus) affects

IOP Conf. Series: Journal of Physics: Conf. Series **1076** (2018) 012006 doi:10.1088/1742-6596/1076/1/012006

the microscale (cells) and the macroscale (human) (image 2). This taught content is associated with the aim PTs to understand that "the nanoscale may impact our everyday world (lesson 5) which is considered as a key concept in NSE education (Delgado *et al.* 2015). Concerning the size dependent properties, PTs related the mechanism of the gecko adhesion with applications in everyday life, such as gloves.

5. Discussion

We consider that certain fundamental NSE concepts have the potential to form a suitable content structure for PTs. "Size & Scale" and "tools" may help teachers to realize the realm of the nanoscale, "applications" provide examples about how the nanoscale can affect our macroscale, "size dependent properties" may be very motivating due to the unexpected behavior of matter at the nanoscale and "models" can bridge the gap between the nanoscale and the macroscale. The above justifications are encountered in similar approaches reported in the literature (e.g. Jones *et al.* 2015).

Table 1. Essential concepts and content structure per lesson

Lesson	Essential concepts	Content Structure	The aim was PTS to
1&2	Size & Scale Tools	• 5 levels of thinking • objects and tools of the macro- micro- nano- scale	• define the nanoscale by its size range, the landmark objects that includes, the tools that render the objects visible
3	Size & Scale Tools Models	• Macroscale • PTs create and evaluate their own models • Nature and role of models	• realize that the macroscale includes objects with different properties • acknowledge that the naked eye is the tool for studying macroscale objects • understand that models represent properties of macroscale objects
4	Size & Scale Tools Models	• Microscale • PTs create and evaluate their own models • Nature and role of models	• realize that the microscale includes objects with different properties • acknowledge that the optical microscope is the tool for viewing microscale objects • understand that models can represent properties of microscale objects
5	Size & Scale Tools Models Applications	• Nanoscale • PTs create and evaluate their own models • Nature and role of models: representations of nanoscale entities • The viral infection case	• realize that the nanoscale includes objects with different properties • acknowledge that electron microscopes can be used for viewing nanoscale objects • realize that models can be used to obtain information about inaccessible targets • understand that the nanoscale may impact our everyday world
6 & 7	Applications Size & Scale Models Size dependent properties	• Lotus effect – applications • PTs create and evaluate their own models • The less the surface contact area between surfaces, the more decreased the adhesion	• understand the super hydrophobic and self-cleaning property of the lotus leaf and the importance of the surface contact area • understand that the nanoscale may impact our everyday world • enhance their understanding that models represent aspects of nanoscale phenomena
8	Applications	• Gecko effect – applications	• understand the strong adhesion property

	Size & Scale Models Size dependent properties	• PTs evaluate models • The more the surface contact area between surfaces, the more increased will be the adhesion	of the gecko lizard and the importance of the surface contact area • understand that the nanoscale may impact our everyday world • refine their understanding that models represent aspects of nanoscale phenomena
9	Size & Scale Models Applications	• Water filtration • PTs create and evaluate their own models	• refine their knowledge about size & scale, models and the impact of nanoscale to macroscale

To avoid the emergence of misconceptions and to enhance NSE content learning, not only we encouraged PTs to use models but also we addressed several aspects of the nature of models in parallel with the NSE content. In addition, the lab work we set consisted of a lot of model building activities. Despite that there are some NSE teacher's training courses that implement model building practices (e.g. Bryan *et al.* 2012), it seems that the interlacement of the NSE content with the content of the nature of models, is an approach that is underestimated in the related literature. On the contrary, we are strongly aligned with the relevant Science Education suggestions which highlight the fact that introducing learners, aspects about the nature of models and modeling can foster learning from models (Oh & Oh 2011).

Some first remarks regarding the effectiveness of the approach reveal that PTs showed a significant deal of enthusiasm and anxiety to participate in the lab activities we designed. They argued that these activities could be suitable for their classes also, with some alternations in order for their students to develop understanding. Moreover, they willingly suggested designing their own experimental activities that would be implemented to their classes. From our point of view, our primary focus is to examine the PTs level of understanding regarding the taught NSE content. The results will point out the modifications to the content structure and to the experimental activities as well.

References

Bhushan B. (2016) Introduction to Nanotechnology: History, Status, and Importance of Nanoscience and Nanotechnology Education. In *Global Perspectives of Nanoscience and Engineering Education, Science Policy Reports* (Winkelmann K. & Bhushan B., eds.) Switzerland: Springer International Publishing, pp. 1-31.

Blonder R. & Sakhnini S. (2016) What Are the Basic Concepts of Nanoscale Science and Technology (NST) that Should Be Included in NST Educational Programs? In *Global Perspectives of Nanoscience and Engineering Education, Science Policy Reports* (Winkelmann K., Bhushan B. eds.), Switzerland: Springer International Publishing.

Bryan L.A., Sederberg D., Daly S., Sears D. & Giordano N. (2012) Facilitating teachers' development of nanoscale science, engineering, and technology content knowledge. *Nanotechnology Reviews*, **1**(1), 85-95.

Daly S. & Bryan L.A. (2010) Model use choices of secondary teachers in nanoscale science and engineering education. *Journal of Nano Education* **2**, 76-90.

Delgado C., Stevens S.Y., Shin N. & Krajcik J. (2015) A middle school instructional unit for size and scale contextualized in nanotechnology. *Nanotechnology Reviews* **4**(1), 51-69.

Duit R., Gropengießer H., Kattmann U., Komorek M. & Parchmann I. (2012) The Model of Educational Reconstruction Framework for Improving Teaching and Learning Science 1. In *Science education research and practice in Europe.* Sense Publishers, pp. 13-37.

Feather J.L. & Aznar M.F. (2011) *Nanoscience Education, Workforce Training, and K-12 Resources.* Boca Raton, FL: Taylor & Francis Group.

Healy N. (2009) Why Nano Education? *Journal of Nano Education* **1**(1), 6-7.

Hingant B. & Albe V. (2010). Nanosciences and nanotechnologies learning and teaching in secondary education: A review of literature. *Studies in Science Education* **46**(2), 121-152.

Hochella M.F. (2002) Nanoscience and technology: the next revolution in the Earth sciences. *Earth and Planetary Science Letters* **203**(2), 593-605.

Huang C.Y., Hsu L.R. & Chen H.C. (2011) A study on the core concepts of nanotechnology for the elementary school. *Journal of National Taichung University: Mathematics, Science & Technology* **25**, 1-22.

Jones G., Blonder R., Gardner G., Albe V., Falvo M. & Chevrier J. (2013) Nanotechnology and Nanoscale Science: Educational challenges. *International Journal of Science Education* **35**(9), 1490-1512.

Jones M.G., Gardner G.E., Falvo M. & Taylor A. (2015) Precollege nanotechnology education: a different kind of thinking. *Nanotechnology Reviews* **4**(1), 117-127.

Kähkönen A.L., Laherto A. & Lindell A. (2011) Intrinsic and Extrinsic Barriers to Teaching Nanoscale Science: Finnish Teachers' Perspectives, *Journal of Nano Education* **3**(1-2), 1-12.

Laherto A. (2011) Incorporating nanoscale science and technology into secondary school curriculum: Views of nano-trained science teachers. *Nordic Studies in Science Education* **7**(2), 126-139.

Lin S.Y., Wu M.T., Cho Y.I. & Chen H.H. (2015) The effectiveness of a popular science promotion program on nanotechnology for elementary school students in I-Lan City. *Research in Science & Technological Education* **33**(1), 1-16.

Logothetidis S. (2012) Nanotechnology: Principles and Applications. In *Nanostructured Materials and Their Applications* (Logothetidis S. eds.), Berlin–Heidelberg Springer-Verlag, pp. 1-22.

Magana A.J., Brophy S.P. & Bryan L.A. (2012) An integrated knowledge framework to characterize and scaffold size and scale cognition (FS2C). *International Journal of Science Education* **34**(14), 2181-2203.

Murty B., Shankar P., Raj B., Rath B.B. & Murday J. (2013) *Textbook of Nanoscience and Nanotechnology*. Universities Press (India) Private Limited.

Oh P.S. & Oh S.J. (2011) What Teachers of Science Need to Know about Models: An overview, *International Journal of Science Education* **33**(8), 1109-1130.

Osborne J. (2008) Engaging young people with science: does science education need a new vision? *School Science Review* **89**(328), 67.

Ramsden J. (2009) *Applied Nanotechnology*. Oxford Elsevier Inc.

Ramsden J.J. & Freeman J. (2009) The nanoscale, *Nanotechnology Perceptions* **5**, 3-25.

Sakhnini S. & Blonder R. (2015) Essential Concepts of Nanoscale Science and Technology for High School Students Based on a Delphi Study by the Expert Community. *International Journal of Science Education* **37**(11), 1699-1738.

Sakhnini S. & Blonder R. (2016) Nanotechnology applications as a context for teaching the essential concepts of NST. *International Journal of Science Education* **38**(3), 521-538.

Schonborn K.J., Host G.E. & Palmerius K.E.L. (2016) Interactive Visualization for Learning and Teaching Nanoscience and Nanotechnology. In *Global Perspectives of Nanoscience and Engineering Education* (Winkelmann K., Bhushan B. eds.) Switzerland Springer International Publishing, pp. 195-222.

Spyrtou A., Zoupidis A. & Kariotoglou P. (2008) The design and development of an ICT- Enhanced Module concerning density as a property of materials applied in floating-sinking phenomena. In *Girep International Conference, Physics Curriculum Design, Development and Validation* (Constantinou C.P. & Papadouris N., eds.), Selected Papers, 391-407.

Stevens S., Sutherland L.M. & Krajcik J.S. (2009) *The big ideas of nanoscale science and engineering: A guidebook for secondary teachers*. Arlington, VA: NSTA Press.

Swarat S., Light G., Park E.J. & Drane D. (2009) Unpacking Student Conceptions of Surface A rea to Volume Ratio in the Nanoscience Context: An Empirical Application of the Construct-Centered Design Framework. In *Proceedings of the Research in Engineering Education Symposium*, Palm Cove, QLD, pp. 273-279.

IOP Conf. Series: Journal of Physics: Conf. Series **1076** (2018) 012006 doi:10.1088/1742-6596/1076/1/012006

Tretter T., Jones M., Andre T., Negishi A. & Minogue J. (2006) Conceptual boundaries and distances: Students' and experts' concepts of the scale of scientific phenomena. *Journal of Research in Science Teaching* **43**(3), 282-319.

Turkoglu A.Y. & Oztekin C. (2016) Science teacher candidates' perceptions about roles and nature of scientific models. *Research in Science & Technological Education* **34**(2), 219-236.

Wansom S., Mason T.O., Hershman M.C., Drane D., Light G., Cormia R., Stevens S. & Bodner G. (2009) A rubric for post-secondary degree programs in nanoscience and nanotechnology. *International Journal of Engineering Education* **25**(3), 615-627.

Xie C. & Pallant A. (2011) The Molecular Workbench Software: An Innovative Dynamic Modeling Tool for Nanoscience Education. In *Models and Modeling Cognitive Tools for Scientific Enquiry* (Khine M.S. & Saleh I.M., eds.), Springer Science + Business Media B.V., pp. 121-139.

IOP Conf. Series: Journal of Physics: Conf. Series **1076** (2018) 012007 doi:10.1088/1742-6596/1076/1/012007

An inquiry-based learning path to introduce modern physics in high-school

Agnese Russo[1], Dominique Persano Adorno[2]

[1]Liceo Scientifico "Benedetto Croce", Palermo, Italy
[2]PERG, Dipartimento di Fisica e Chimica, Università di Palermo, Italy

Abstract. Although Modern Physics (MP) is a scientific field of increasingly interest, due to its mathematical and conceptual difficulty, it was not generally taught in Italian high-schools. Recently, in order to reduce the gap between school curricula and modern technology, the Ministry of Education promoted the inclusion of MP in Italian secondary schools. MP hot topics produce interest and enthusiasm in learners at all ages, but to address this topic in high school in an effective way it is necessary to translate complex theories and highly-advanced experiments into an understandable language appreciated by students. In this contribution, we discuss an inquiry-driven learning sequence suitable to introduce MP, that we experienced, as pilot study, with a sample of about 40 students attending the last year of the scientific high school "Benedetto Croce" of Palermo in the school year 2015/2016. Our results suggest that an inquiry-based learning path, composed by three phases (frontal lessons + inquiry-based laboratorial activity + cooperative work), can constitute a successfully teaching approach to effectively engage students into an active learning of the MP, provided the teacher has a crucial role in facilitating the knowledge, activating the questioning process and supporting a valuable reasoned exploration.

1. Introduction

The importance of teaching Modern Physics (MP) in secondary school isn't just connected to its cultural relevance. In fact, this surprising and charming branch of Physics, builds a new way of thinking and seeing the world. MP is not only an ensemble of calculation rules to reproduce experimental data; its objective is to develop new technologies that have a ubiquitous presence in everyday life. Moreover, MP offers the possibility of building concepts and interpretative hypothesis, which haven't their counterparts in classical area. Because of this, the Italian Ministry of Education recently decided to include it in the curricula of secondary school. However, some problematic aspects and obstacles to overcome arise both in the teaching and in the learning of the MP (Persano Adorno *et al.* 2017). In the last decades, physics educators proposed many teaching strategies to better address this issue. There are several important studies focused on the learning and teaching MP at secondary school and senior/junior college levels (Kroemer 1994, Styer 1996, Petri & Niedderer 1998, Johnson *et al.* 1998, Ireson 2000, Taber 2004, Persano Adorno & Pizzolato 2015). Many researchers concluded that both teaching and learning MP are hard because it contains abstract ideas, requires strong mathematical tools, and deals with complicated operations. In secondary school environments, most problems deal with the teaching strategies adopted for MP and conceptual difficulties (Kroemer 1994, Ireson 2000, Muller & Wiesner 2001, Olsen 2002, Etkina 2010).

To effectively introduce MP at the high-school level, teachers need to translate complex theories and highly-advanced experiments into an understandable language appreciated by students. Further-

IOP Conf. Series: Journal of Physics: Conf. Series **1076** (2018) 012007 doi:10.1088/1742-6596/1076/1/012007

more, it is increasingly necessary that educators change their attitude towards this discipline, renouncing a too transmissive teaching method. Another relevant problem in the teaching/learning of MP is related to the low level of textbooks, very often unsuitable for the learners; although they introduce new concepts in a simple way, they appear too much detailed about particular aspects. The textbooks look as encyclopedias, built as university manuals. Finally, the choice of arguments without taking account of psychological maturation of students, the excessive overexposure of MP topics on Internet, the underestimation of historical and philosophical aspects of the discipline, are all elements limiting the acquisition of skills by students.

In order to present MP, the teacher can make operational choices between different educational approaches: (i) a *Technological* approach, based on main applications of MP: laser, diodes, sheet of graphene, etc.; (ii) a *Philosophical* approach, mainly based on the interpretation of MP and on its meaning for the worldview; (iii) a *Historical* approach, based on the dealing with the problems of classical physics solved by MP, (iv) an *Experimental* approach, based on performing experiments in the labs in cooperative learning; (v) an *Inquiry* approach, based on survey scientific approach through an active exploration process, or a suitable combination of them.

On the basis of our teaching experience, we believe that the analysis of the phenomenology, the implementation of experimental activities, the modelling and the use of data representation languages, the contextualization of theoretical concepts in the world where we live, are important elements to facilitate the building of learning of MP. In the following we present and discuss an inquiry-driven learning sequence suitable to introduce MP, experienced by students attending the last year of the lyceum "Benedetto Croce" of Palermo. We show that an inquiry-based learning path, composed by three phases (frontal lessons + inquiry-based laboratorial activity + cooperative work), can constitute a successfully teaching approach to effectively engage students into an active learning of the MP, provided the teacher has a crucial role in facilitating the knowledge, activating the questioning process and supporting a valuable reasoned exploration.

2. A brief introduction to an inquiry-based activity

The purpose of an inquiry-based process is to make learners able to analyze a problem, to run and develop a plan of studies and exploration activities, to build speculations, by making a distinction between possible alternatives, to search for information, to build models, to discuss and compare them in a context of peers (Herron 1971, Bybee 1993, McDermott 1996, National Research Council 2000, Llewellyn 2002, Banchi & Bell 2008). The Inquiry-based approach also allows to relate existing knowledge and new experiences, to change preconceived ideas and conceptual design, and to build new knowledge. In particular, students involved in an inquiry-based learning activity, collect data to develop explanations of scientific phenomena they study. They give a particular importance to the experimental evidence, using it as a starting point to build explanations on the behavior of natural phenomena; they build models designed to answer scientific questions and, doing it, build high-level skills; they discuss and communicate, in a context of peers, their models of explanations, developing critical review skills, crucial in their future life.

The teachers play the delicate role of planning a scientific inquiry program for their students by selecting approaches able to strengthen the understanding in students, choosing contents and contexts that create interests. They facilitate the learning, focusing on the scientific inquiry, organizing working groups, challenging the students to accept and share responsibilities related to their learning, identifying the differences among students and promoting these. They activate continuous assessment strategies of their teaching and student learning, steadily checking the understanding, directing the self-assessment, improving the teaching practice. The teachers design the learning environments, define the working times, identify useful resources outside of school and endeavor to make these resources available to students. Educators promote in their classes the development of scientific communities dedicated to learning, highlighting different ideas, skills and experiences of learners and facilitating a fruitful collaboration among students. Fig. 1 shows the different phases of an Inquiry-based learning cycle.

IOP Conf. Series: Journal of Physics: Conf. Series **1076** (2018) 012007 doi:10.1088/1742-6596/1076/1/012007

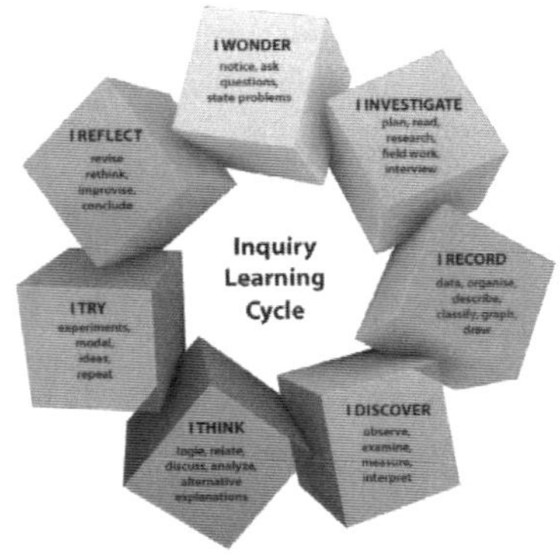

Fig. 1. Inquiry-based learning cycle (http://www.creativeinquiryteacher.com/?page_id=6)

There are various levels of inquiry in science education – the initial level where the teacher directs every aspect, to the highest level where the student holds the control and needs the intellectual and practical skills to become investigator, acting as a researcher (Banchi & Bell 2008). Through the series of different levels of inquiry, the student becomes more able to carry out his/her own independent inquiry, and the assistance of the teacher becomes different, less instructive, but more enabling and flexible.

3. Activity description

Our sample of students attending the last year of a scientific high-school actively participated to a cooperative learning on the fundamental concepts of MP and was involved in a laboratorial activity concerning the Franck-Hertz and the photoelectric effect experiments within an inquiry-based learning environment, with different level of teacher guidance.

Indeed, before the beginning of the laboratorial work, we randomly split up our sample in two groups composed by about 20 students:

Group 1 faced the experimental activity in a ***Guided-Inquiry*** learning environment (Banchi & Bell 2008). In this situation, it is expected that the students plan and conduct the experiment, with little guidance from the teacher and with a limited laboratory preparation. The research problem to be solved is given by the teacher, but the students have the responsibility to design and conduct the work, by collecting data and building models.

Group 2 faced the laboratorial work within a ***Structured-Inquiry*** environment (Banchi & Bell 2008). In this level of Inquiry, the question and the detailed procedure for the utilization of the labs are provided by the teacher. However, the students, working in small groups, generate an explanation supported by the collected data by themselves in the lab. They are responsible for uncovering the answer. The teacher acts as a knowledge facilitator, providing support or materials.

The whole activity, lasting about 15-20 hours, consisted of following three phases:

Frontal lessons, mainly based on video-projection and use of multimedia material, tests and oral discussion;

1. Execution of experiments at the Modern Physics Laboratory at the Department of Physics and Chemistry of the Palermo University;
2. Cooperative work in the classroom, which included the data analysis and the conclusive discussion.

In these phases, great importance was done to implement peer discussions in order to favour problem-solving among students.

Phase 1

During the first phase of the learning path, the students of both groups were introduced by the educators into the physical problem to address in the Franck-Hertz and in the Photoelectric effect experi-

IOP Conf. Series: Journal of Physics: Conf. Series **1076** (2018) 012007 doi:10.1088/1742-6596/1076/1/012007

ments. In this phase, all students actively participated to the discussion with questions focused on the limits of classical physics, on the Bohr model and on the Einstein explanation. This phase lasted 4 hours for both groups.

Fig. 2. A view of the frontal lesson (*Phase 1*) at the scientific high school "B. Croce" of Palermo

Phase 2

Before the **laboratorial activity**, at the Modern Physics Laboratory, the students were invited to follow the educators' instructions for what concerns the procedure to perform, in order to get to the necessary physical conditions for the data collection. Fig. 3 shows the experimental facilities available at the Laboratory for both the experiences.

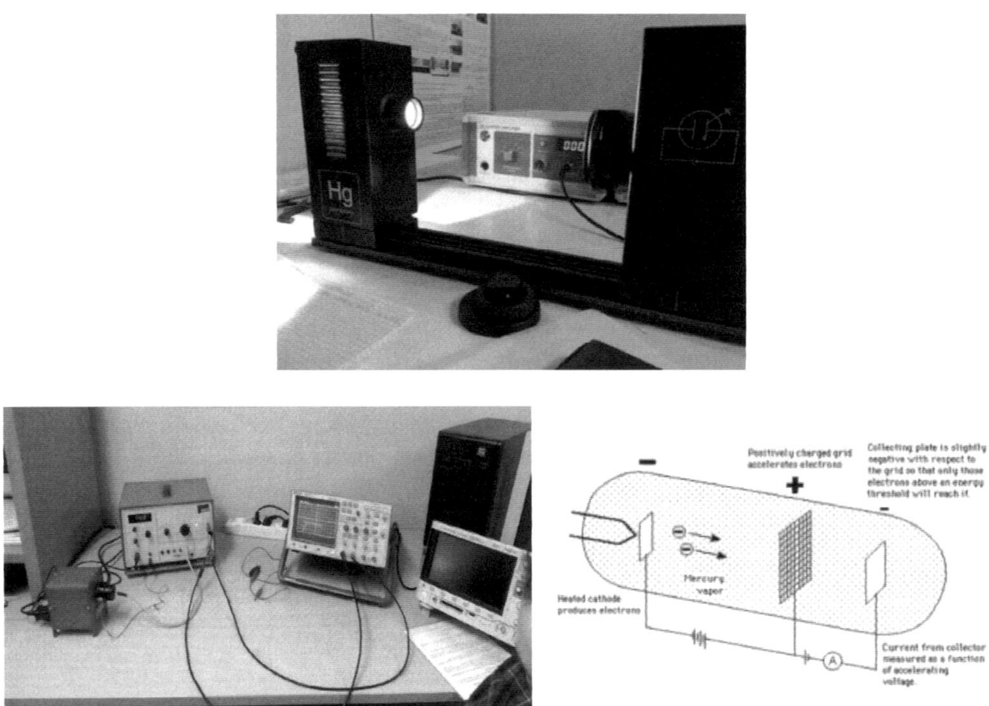

Fig. 3. (top): Experimental set-up for the Photoelectric effect; (bottom): Experimental set-up for the Franck-Hertz experiment (left) and schematic representation of the mercury-filled F-H tube (right)

IOP Conf. Series: Journal of Physics: Conf. Series **1076** (2018) 012007 doi:10.1088/1742-6596/1076/1/012007

Group 1

Group 1 was firstly invited to carry out the Photoelectric Effect experiment. The teacher, showed the tool use and proposed the research question: "Where and why the classical physics fails in the data explication?". In order to answer to this question, the students had to develop hypotheses, collect data and draw conclusions based on the recorded data.

The first afternoon (three hours) in the laboratory were critical. Students belonging to this group seemed quite bewildered and stuck on a stance. In this phase, the teacher tried, by using more questions, to stimulate and encourage them to go forward (Persano Adorno & Pizzolato 2015). For example, despite students well knew that they were dealing with a photodiode, in accordance with their previous knowledge, they tried to verify the first Ohm's law.

In order to put an end to the sense of failure experienced by students of group 1 when working at this level, during the second afternoon (three hours) spent in the laboratory, the teachers decided to actively participate to the experiment conduction, illustrated the procedure and the method and provided the results and their explanation, lowering the level of the Inquiry approach towards a Confirmation Inquiry environment (Banchi & Bell 2008). Despite of the students' diligence on performing the scheduled measurements and on collecting the data, they encountered many difficulties to generate their own explanations.

Group 2

After a more detailed instruction on the question and on the utilization of the experimental tools, students of group 2 were asked to work in groups and to perform scientific investigations, by collecting data and building possible explanations. The active participation of the teachers to the discussion, as peers, activated student scientific inquiry through the onset of an effective questioning on what was observed and measured.

Students used the logbooks to annotate the followed procedure, the difficulties encountered throughout the activity and the changes they made during the inquiry process. The students of this group spent ten hours in the lab successfully carrying out both the experiences.

Fig. 4. Students carrying out the Photoelectric effect

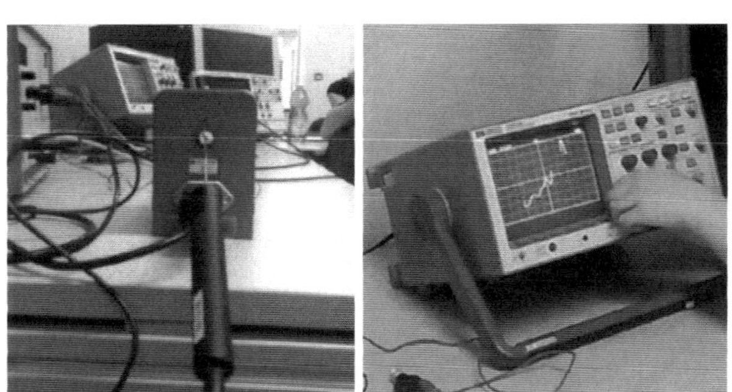

Fig. 5. Students carrying out the Franck-Hertz experiment

Phase 3

The **cooperative work** was carried out in the classroom, lasting four-five hours. Students of both groups were invited to write a scientific report on the experience done and on acquired/reinforced concepts.

In particular, students of **Group 2** built the models and performed the calculations, drawing the graphs for both the experiences. By starting from the data collected during the Franck-Hertz experiment, they were able to find an experimental confirmation of the Bohr's postulates, asserting that atoms can absorb energy only in quantum portions. During the cooperative discussion in the classroom they argued that in this experiment, electrons are accelerated and pass through mercury vapor, where they lose energy by inelastic scattering in quantized steps as they excite mercury atoms from the ground state to an excited state. By measuring the distance between the peaks in the I-V curve, they found that the excitation energy is $E_{exc} \sim 5eV$, in very good agreement with the real value of the difference in energy between the two levels in mercury ($\Delta E = 4.9$ eV).

For what concerns the photoelectric effect, the students belonging to Group 2 carried out the best fit of the data collected (V_{stop}-vs-ν) in order to calculate the value of the Planck's constant. During this relevant phase of the cooperative work, the students learned the significance of linear regression of the experimental data and how to calculate the error bars. They obtained the very reasonable value: $h = (6.51 \pm 0.25)10^{-34}$ Js.

Unfortunately, both for sake of time and for lack of interest, students belonging to **Group 1** did not carry out the Franck-Hertz experiment; for what concerns the photoelectric effect, they did not find the relationship between the potential stop and the frequency values and limited themselves to write a report, without data analysis and calculation of the Planck's constant.

Fig. 6. Cooperative work (*Phase 3*) of students in the classroom

4. Discussion and conclusion

Information about the students belonging to both groups affective development and motivation to learn was achieved by means of structured interviews based on the Intrinsic Motivation Inventory, with specific items adapted to our study (McAuley *et al.* 1991, Ryan & Deci 2000, Jang *et al.* 2016). To quantify the student satisfaction, we used a five-point Likert scale: 5: Very much; 4: Somewhat; 3: Undecided; 2: Not really; 1: Not at all.

Table 1. Mean student outcomes on a five-point Likert scale

Questions (interest-enjoyment dimension; perceived competence dimension)	Confirmation Inquiry Group 1	Structured Inquiry Group 2
I enjoyed this learning experience	2.9	4.5
I am satisfied with my performance at this experience	2.1	4.4
After this learning experience, I feel pretty competent	1.8	3.9
I understood better the topic	1.5	4.5

As evidenced also by the answers to the interviews, the students of **Group 1** were not able to cope the sense of responsibility of which were been charged and, during the laboratorial learning activity, their attitude was passive and unmotivated. The sense of failure generated frustration in the students, making them consider the laboratorial work "a waste of time" (Quintana *et al.* 2005).

Otherwise, the laboratorial learning sequence followed by students of **Group 2** enhanced their outcomes in terms of motivation, interest, reflective participation, critical observation and discovery learning. The learners have enthusiastically realized a video testifying the whole activity. Moreover, some students of Group 2, discussed this activity at their final graduation exam.

In conclusion, the objective of the learning path here described was not to add extra contents to those delivered during regular classes, but try to modify the student conceptions on MP. Indeed, in previous years we noticed that traditional lectures and discussions, culminating in the resolution of some exercises, many times did not encompass quantum concepts that remain unclear.

The close synergy between high school and university educators represented a positive experience, as well as the laboratorial work. On the basis of (i) the answers to structured interviews, (ii) the marks obtained at the final examination, our results suggest that an inquiry-based learning path, composed by three phases (frontal lessons + inquiry-based laboratorial activity + cooperative work), can constitute a successfully teaching approach to effectively engage students into an active learning of the MP, provided the teacher has a crucial role in facilitating the knowledge, activating the questioning process and supporting a valuable reasoned exploration.

Inquiry-based learning environments with lower teacher guidance, helpful to stimulate higher reasoning skills, particularly in university context (Pizzolato *et al.* 2014), sometimes may produce negative feelings due, for example, to run into mistakes or achieve unexpected results, especially in high-school students.

References

Banchi H. & Bell R. (2008) The many levels of inquiry, *Science and Children* **46**(2), 26-29.

Bybee R.W. (1993) An instructional model for science education. In *Developing Biological Literacy*, Biological Sciences Curriculum Study, Colorado Springs, CO.

Etkina E. (2010) Pedagogical content knowledge and preparation of high school physics teachers, *Physical Review Special Topics Physics Education Research* **6**, 020110.

Herron M.D. (1971) The nature of scientific enquiry, *School Review* **79**, 171-212.

Ireson G. (2000) The quantum understanding of pre-university physics students, *Physics Education* **35**, 15-21.

Jang H., Reeve J. & Halusic M. (2016) A New Autonomy-Supportive Way of Teaching That Increases Conceptual Learning: Teaching in Students' Preferred Ways, *The Journal of Experimental Education* **84**(4), 686-701.

Johnson I.D, Crawford K. & Fletcher P.R. (1998) Student difficulties in learning quantum mechanics, *International Journal of Science Education* **20**, 427-446.

Kroemer A. (1994) Investigating quantum physics in the high schools environments, *Physics Education* **56**, 127-139.

Llewellyn D. (2002) *Inquiry Within: Implementing Inquiry-based Science Standards*, Corwin Press Inc., Thousand Oaks, CA.

McAuley E., Wraith S. & Duncan T.E. (1991) Self-efficacy, perceptions of success, and intrinsic motivation for exercise, *Journal of Applied Social Psychology* **21**, 139-155.

McDermott L.C. & Physics Education Group at the University of Washington (1996) *Physics by Inquiry*, John Wiley & Sons, New York.

Muller R. & Wiesner H. (2001) Teaching quantum mechanics on an introductory level, *Am. J. Phys.* 70, 200-209.

National Research Council (NRC). (2000), *Inquiry and the National Science Education Standards: A Guide for Teaching and Learning*, The National Academies Press, Washington DC.

Olsen R.V. (2002) Introducing quantum mechanics in the upper secondary school: a study in Norway, *International Journal of Science Education* **24**, 565-574.

Petri P. & Niedderer H. (1998) A learning pathway in high-school level quantum atomic physics, *International Journal of Science Education* **20**, 329-347.

Persano Adorno D. & Pizzolato N. (2015) An inquiry-based approach to the Franck-Hertz experiment. *Il Nuovo Cimento C* **38**, 109.

Persano Adorno D., Fazio C., Pizzolato N. & Battaglia O.R. (2017) Training Pre-service and In-service Secondary School Teachers: Analysis of Changes in Perceptions About Quantum Physics Concepts and NoS Views. *Key Competences in Physics Teaching and Learning* **190**, Springer Proceedings in Physics, pp. 165-176.

Pizzolato N., Fazio C., Sperandeo-Mineo R.M. & Persano Adorno D. (2014) Open-inquiry driven overcoming of epistemological difficulties in engineering undergraduates: a case study in the context of thermal science", *Physical Review Special Topics Physics Education Research* **10**, 010107.

Quintana C., Zhang X. & Krajcik J. (2005) A framework for supporting meta cognitive aspects of online inquiry through software-based scaffolding, *Journal of Educational Psychology* **40**, 235-244.

Ryan R.M. & Deci E.L. (2000) Self-determination theory and the facilitation of intrinsic motivation, social development, and well being, *American Psychologist* **55**, 68-78.

Styer D.F. (1996) Common misconceptions regarding quantum mechanics, *American Journal of Physics* **64**, 31-34.

Taber K.S. (2004) Learning Quanta: barriers to stimulating transitions in student understanding of orbital ideas, *Science Education* **89**, 94-116.

IOP Conf. Series: Journal of Physics: Conf. Series **1076** (2018) 012008 doi:10.1088/1742-6596/1076/1/012008

Low cost experimental proposals to bridge from classical to modern physics

Daniele Buongiorno, Giuseppe Fera, Marisa Michelini

Università degli Studi di Udine

Abstract. Simple experiments with diffraction gratings allow us to build a bridge between classical and modern physics. A LED, a diffraction grating and a screen are sufficient to measure the wavelength of the light emitted by the LED. The energy E of the emitted photon is inversely proportional to wavelength of the light λ, $E = hc/\lambda$, where h is Planck's constant. By a simple multimeter and a volt-ampere connection we can obtain the characteristics voltage-current curves of the LED. The threshold voltages of LEDs of different colours are easily measurable. When the LED emits light, we can assume that the electron of the current circulating in the LED when the voltage U transfers its energy eU to the photon and we can derive the Planck's constant by the relationship $h = \lambda eU/c$. The students of the Quadri Lyceum of Vicenza have successfully tested such an activity within a course on electrical conduction. With the data for the red LED the students obtain $h = 6*10^{-34}$ Js. Now, by means of a CD-ROM we build a simple device to carry out the same measurement in reflection modality. The use of a simple and economic spectroscope enriches the experimental proposal in the perspective to build a path bridging from classical to modern physics. The intervention module based on this kind of lab in modern physics summer school for secondary students allows us to analyze its role in learning.

1. Introduction

Simple and quantitative experiments are important didactic activities that contribute to the learning process in two ways (Viennot *et al.* 2014) concerning:

- the specific contents;
- the methodological aspects regarding physics as a discipline (Holbrook & Rannikmae 2007).

These experiments give an even more significant contribution if they concern modern physics topics, employing high-tech devices, nowadays of common use: in this perspective, LEDs represent a unique context to develop lots of proposal (Planinšič & Etkina 2014). With this cheap and common piece of equipment, it is possible to treat, at different levels, various problems concerning electrical conductivity in solids, as well as the ones concerning physical optics or spectroscopy. In particular, optical spectroscopy represents a bridge between classical and modern physics, and it has been recently subject of research as concern learning processes (Ivanjek *et al.* 2015a, 2015b). Moreover, the study of optoelectronic properties of LEDs represents a fertile link with material physics, that student can directly deal with. These potentialities have been tested in a pilot proposal conducted in 2015 by professors G. Fera & D. Merlin, that inspired a new laboratory-based proposal, used as a didactical intervention in 2016.

2. The pilot proposal and the basis of our proposal

The pilot proposal and the learning results

A teaching intervention module on electrical conductivity in solids was carried out in 2015 at Lyceum Quadri of Vicenza (IT) with students aged 18-19 by means of IBL strategy to correlate macro quantities (U-I-R) with microscopic processes by means of Drude's model. An experimental problem solving was used to face the interpretative problem of the emission of light by a LED to bridge from classical to quantum model (energy bands). Characteristics curves U-I are measured for different LEDs (Fig. 1) in order to correlate the threshold voltage U to the frequency ν or the wavelength λ of the emitted light, measured with the apparatus described in Fig. 2.

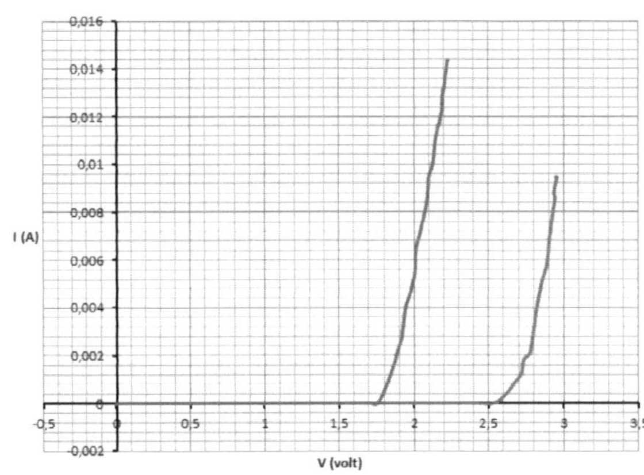

Fig. 1. U-I characteristics of a red and a blue LED

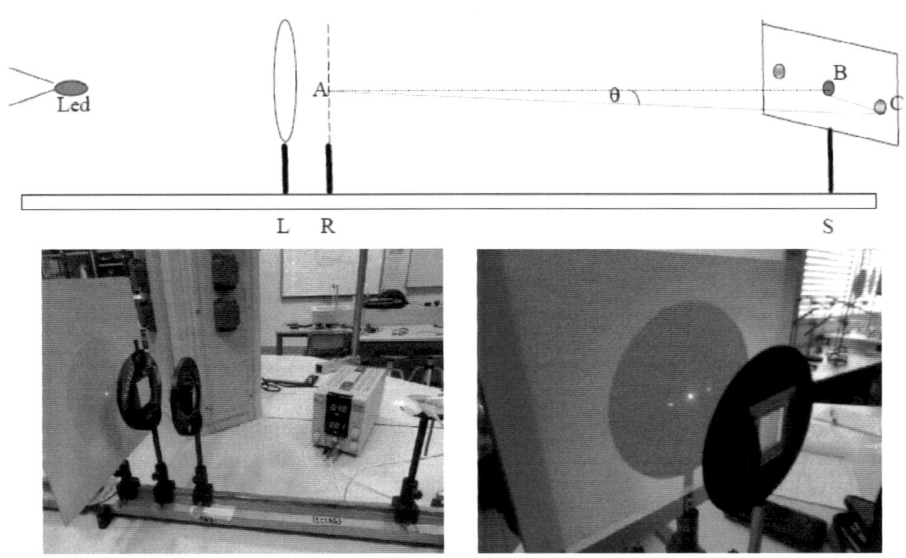

Fig. 2. Measurement of wavelength: experimental setup. L is the lens, used to make the light collimated, R is the diffraction grating, and S the screen

Data obtained according to the table 1 allow an evaluation of the Planck's constant h.

IOP Conf. Series: Journal of Physics: Conf. Series **1076** (2018) 012008 doi:10.1088/1742-6596/1076/1/012008

Table 1. Data acquired by students in order to evaluate Planck's constant h

θ (rad)	d (μm)	λ = d sinθ (μm)	ν = c/λ(s^{-1})	U(V)	E = eU (eV)	h = eUλ/c (Js)
0.38	1.75	0.65	4.62*10^{14}	1.75	1.75	6.07*10^{-34}

The monitoring of students' reasoning evidenced the following aspects:

- Drude's model is no more valid for the interpretation of the emission of a LED (10/11) because: LED is a semiconductor (7/11); there are no metallic bonds (2/11); LED is not a metal (1/11);
- Activation energy is correlated to the color of the LED (6/11); in order to produce a current, electrons have to overcome a limit (1/11);
- The activation energy is inversely proportional to the wavelength (3/11);
- The relation between energy and frequency of the emitted light is supposed to be a direct proportionality (5/11).

Our experimental proposal: basis and research questions

Two main needs emerged from this experience:

- To simplify the apparatus to offer each student the opportunity to carry out a personal analysis of the phenomenon;
- To focus on the relationship between energy levels interpretative model and the emitted spectra of a source.

For the first need we developed the two experimental apparatuses (see next section). The former make use of a CD to carry out the measurement in reflection modality, the latter make use of a cheap diffraction grating and a ruler. In both cases the images are virtual. No lenses are needed. For the second goal we used the second version during a summer school (see section 4). Research questions were:

- RQ1: How does experimental activity contribute in identifying: the specific spectral characteristics, the role of a diffraction grating, the relationships between diffracted spectra and energy model of the emitting systems?
- RQ2: Which are the operative and conceptual difficulties encountered by students in the proposed activity?
- RQ3: How do students sketch and compare the energy structure of a gas-discharge lamp (used in another experiment) and of a LED correlating them to the emitted spectra?

3. Experimental proposals

Proposal 1: CD in reflection modality

A CD works as a reflection grating. Placing a light source above its surface (Fig. 3) allows to see the first-order maximum (Fig. 4). In a simplified model (Fig. 5) the angle θ can be determined by geometric measures, and thus $\lambda = d \sin\theta$ ($d = 1.67$ μm for CDs). A potentiometer varies the voltage on the LED; when the LED starts emitting light, it is possible to measure its threshold voltage by means of a voltmeter (Fig. 3). Data concerning red and green LEDs are summarized in table 2.

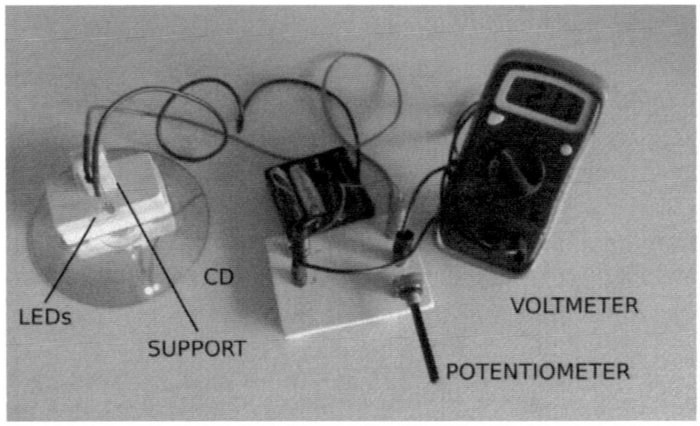

Fig. 3. The prototype used to measure λ of LEDs by means of a CD

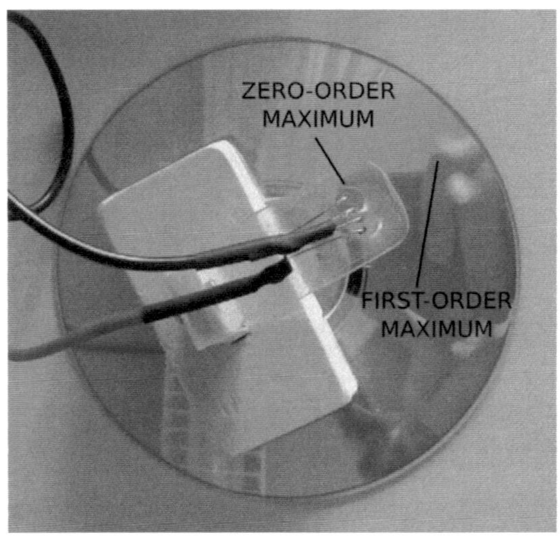

Fig. 4. The CD seen from above

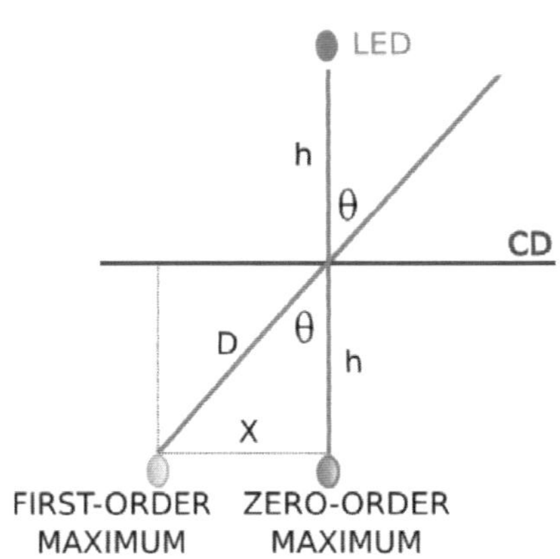

Fig. 5. Oversimplified geometric model for measuring θ

IOP Conf. Series: Journal of Physics: Conf. Series **1076** (2018) 012008 doi:10.1088/1742-6596/1076/1/012008

Table 2. Red and green LEDs: evaluation of λ using a CD and simple geometric measurements[7]

colour	h (cm)	u(h) (cm)	X (cm)	u(X) (cm)	$D = \sqrt{X^2 + h^2}$ (cm)	u(D) (cm)	sinθ = X/D	u(sinθ)	λ = d sinθ (nm)
Red	7.000	0.058	2.800	0.058	7.500	0.058	0.3733	0.0082	653 ± 29
Green	7.000	0.058	2.300	0.058	7.400	0.058	0.3108	0.0082	544 ± 29

Proposal 2: grating and ruler

The setup in Fig. 6 allows students to measure the dominant wavelength emitted by a LED in transmission modality. Looking through cheap rainbow-glasses, used as a diffraction grating, it is possible to see the spectrum of a LED along a ruler. Simple geometric measures allow to obtain the angle θ of the emission peak and thus its wavelength (Figs 6, 7). The implementation of this kind of setup in a didactical proposal is described in the next section.

4. The proposal for the summer school 2016 in Udine

In the 2016 Summer School in Modern Physics at University of Udine (IT) for 32 selected talented secondary school students, two experiments about spectroscopy have been proposed. In the first experiment, students measure wavelengths of a discharge lamp through an optical goniometer, in the second experiment, students see and measure the spectrum of a LED using setup presented in Fig. 6 Learning processes were monitored by means of test-in, test-out, tutorials and interviews. Here we report a case study carried out by means of detailed analysis of the experimental reports written by 4 students and the results coming from two questions from the test-out submitted to the whole group of students.

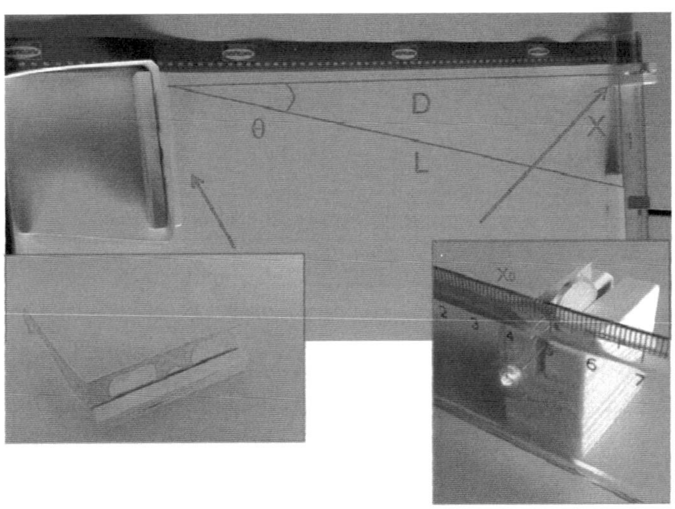

Fig. 6. The low-cost apparatus: with rainbow glasses and a ruler it is possible to see and measure the spectrum of a LED

[7] The combined standard uncertainties $u_c(y)$ of the measurement results y have been evaluated using the formula $u_c(y) = \sqrt{\sum_i \left(\frac{\partial y}{\partial x_i} \cdot u(x_i) \right)^2}$ (the law of propagation of standard uncertainties) where $y(x_i)$ is the evaluated (dependent) quantity and $u(x_1)$ is the standard uncertainty of the measured (independent) variable x_i. The uncertainty of λ is the expanded uncertainty.

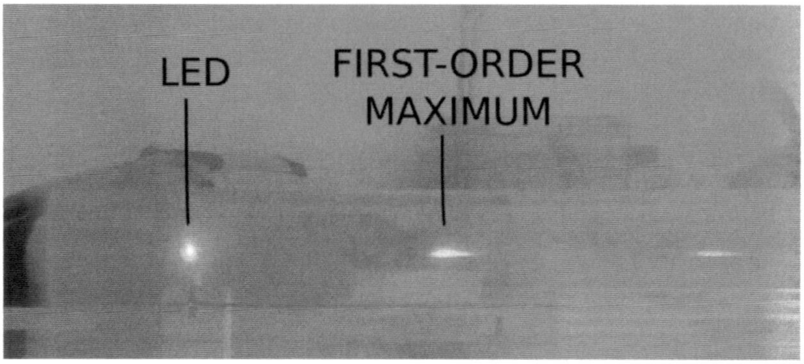

Fig. 7. The grating of the rainbow-glasses allows to see the virtual image of the spectrum of the LED along the ruler

The case study on the lab activity

We analyzed the detailed reports of 4 students concerning this low-cost experiment. We provided them a synthetic sheet for the experimental procedure. Students structured independently the reports in the following sections: introduction, goal, theoretical aspects, material used, procedure, data, discussion and conclusions.

Main learning outcomes

- The energy of the emitted radiation (3 students) is associated to the observed peak (2) or the measured wavelength (1) with specification concerning the energy changes in the emitting system (3) and relative energy level model (1);
- Non-perfect monochromatic nature of the light emitted from a LED (4): the spectrum of a LED consists of a main peak superimposed on a sort of continuum (3), white light spectrum (2) or waves and colors associated (1);
- Intensity decreases with the order, so that the first order is the brightest (2);
- The diffraction grating splits the light (2) in the different wavelength (1);
- Even with simple materials it is possible to obtain reliable measurements of wavelength (3).

Explicit operative difficulties pointed out

- To sight the exact position of the first-order peak;
- To measure the exact LED-grating distance;
- To quantify the uncertainty in measuring λ.

General conclusions: conceptual limits

- The way in which spectral features allow to find the energy levels of the emitting system;
- The way in which students look at the experiment, remaining on observational/operative plans;
- The lack of analysis and argumentations in using formal relationships;
- The relationships between light wave model, energy and colors observed and/or the role of the different components of the apparatus.

Post-test questions submitted to the whole group of students

We analyzed the answers given by 32 students to the following two post-test questions:

Q1) Compare the emission spectra of a LED and a gas-discharge lamp, pointing out similarities and differences.

Q2) Make an hypothesis on the energetic structure of the LED and gas-discharge lamp by means of a sketch.

Q1 – analysis

Similarities pointed out by 12/32 students concern that both sources emit radiation (4) and both have a spectrum (4) that is not continuous (2), polychromatic (5), incomplete (1), with only certain energies (2). The spectra have certain common frequencies/wavelength (2). Differences pointed out by 22/32 students concern that the spectrum of the LED is a continuum (16) or with bands (2) but is monochromatic (4). One student states that the LED needs less energy than the lamp in order to emit. The lamp shows a discrete spectrum (3) characterized by sharp (2) and colored (1) lines (11) due to a superimposition of colors and frequencies (3).

Q2 – analysis

From drawings/statements by 19/32 students, the lamp has a discrete number of energy levels (16) in number of 4 (6) to account for 6 lines (3). Students draw also 3 levels (2) or 5 (2), 6 (2), 7 (1). Representations for the energetic structure of LED fall in 7 models (A-G), represented by the sketches in Fig. 8.

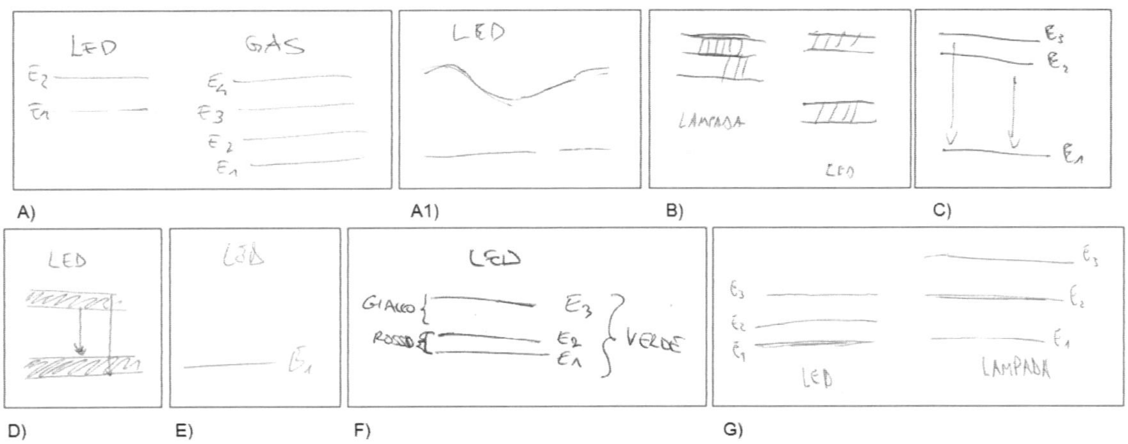

Fig. 8. Representations for the energetic structure of a LED: model A (2 students) and model A1 (1) make use of two energy levels; model B (3) shows transitions between couples of levels; model C (4) represents the left and right edge of the continuous spectrum; model D (3) shows bands; model E (1) makes use of a single level; model F (2) shows all the colors in the spectrum; in model G (1) the energy levels in the LED are more closely spaced than the ones in the lamp

5. Conclusions

Simple experiments, as the ones suggested here, are crucial for conceptual understanding of the relationship between diffracted spectra and energy models. No particular difficulties emerge in stimulating students to switch from a descriptive to an interpretative level. Questions asked are fertile in spontaneous production of energy-levels model for the emitting systems. The richness of the models used by students to describe the energy structure of a LED is a didactical resource as concern the way in which physicists use spectroscopic investigations to build an energy-level model of matter, and the way this knowledge allows to build high-tech devices, as LEDs, to obtain the desired quasi-monochromatic emission.

References

Holbrook J. & Rannikmae M. (2007) The Nature of Science Education for enhancing Scientific Literacy, *International Journal of Science Education* **29**(11), 1347-1362.

Ivanjek L., Shaffer P.S., McDermott L.C., Planinic M. & Veza D. (2015a) Research as a guide for curriculum development: An example from introductory spectroscopy. I. Identifying student difficulties with atomic emission spectra, *American Journal of Physics* **83**(1), 85-90.

Ivanjek L., Shaffer P.S., McDermott L.C., Planinic M. & Veza D. (2015b) Research as a guide for curriculum development: An example from introductory atomic spectroscopy. II. Addressing student difficulties with atomic emission spectra, *American Journal of Physics* **83**(2), 171-178.

Planinšič G. & Etkina E. (2014) Light Emitting Diodes: A hidden treasure, *The Physics Teacher*, **52**(2), 94-99.

Viennot L., Mueller A., Planinšič G. & Sassi E. (2013) Colour phenomena and selective absorption: the role of thickness of absorbing medium, activity sheets, MUSE project of the EPS-PED.

Viennot L. (2014) Thinking in Physics – The pleasure of reasoning and understanding, Springer Ed.

IOP Conf. Series: Journal of Physics: Conf. Series **1076** (2018) 012009 doi:10.1088/1742-6596/1076/1/012009

From led light signboards to the Planck's constant

Francesca Santonocito[1], Antonio Tornabene[1], Dominique Persano-Adorno[2]

[1]Istituto di Istruzione Superiore "E. Fermi - F. Eredia" Catania, Italy
[2]PERG, Dipartimento di Fisica e Chimica, Università di Palermo, Italy

Abstract. Recent studies have highlighted an alarming decline in young people's interest towards the study of scientific subjects, often considered interesting but not easily understood and appreciated by students. In particular, the introduction of Modern Physics (MP) key topics at secondary school level is a difficult and complex task because MP involves abstract ideas and requires a strong mathematical background.
In this communication we present and discuss the results of an inquiry-based teaching/learning path aimed at surmounting the difficulties of an exclusively theoretical approach to the introduction of MP topics. In particular, we planned and realized an inquiry-driven learning environment where about 20 students, from a second-year class of a vocational school, were involved in the discovery of the Planck's constant and brought closer to MP. The work, carried out as a cross-disciplinary module, has involved Physics and Technological Laboratory. The task given to the students was the design of an illuminated sign, efficient from the energetic point of view, flexible and attractive. Through a 'guided inquiry', the students evaluated the different solutions, developed their critical thinking and decided to implement them with LED diodes. Planck's constant has been determined by two methods of analysis of the experimental data and satisfactory values have been obtained. Our results suggest that an inquiry-based teaching/learning path, can constitute a successfully teaching approach to effectively engage students into an active learning of the MP. The described learning activity provided students with opportunities to develop a large range of complementary skills such as working in groups, synthesis, interpretation and evaluation of the experimental data, experience of open-ended problems solving and other cross-disciplinary abilities.

1. Introduction

In the last decades, physics educators proposed many teaching strategies to introduce MP key topics at secondary school level, concluding that both teaching and learning MP concepts is difficult because they contain abstract ideas and require a strong mathematical background (Johnson *et al.* 1998, Kroemer 1994, Ireson 2000, Muller & Wiesner 2001, Olsen 2002, Etkina 2010). Recent reports showed that one of the factors that influence the increase of interest, of motivation and of a positive attitude towards the study of Physics, is represented by the didactic methods used within the teaching-learning process (Global Science Forum 2008).Thus, to the difficulty of teaching MP often adds a wrong teaching methodology implemented by educators, who prefer standard class lectures and mathematical calculations. In order to reach out the teachers and to overcome these problems the Italian Ministry of Education has promoted the development of specific courses for in-service teachers, who should be trained to help the students to acquire scientific and technical skills also in Modern Physics (Pospiech 2000; Michelini *et al.* 2000, 2004a, 2004b; Francaviglia *et al.* 2012). After attending one of these training courses, organized by the Regional Office for Secondary Education, jointly with a team of researchers at the Department of Physics and Chemistry, University of Palermo, within the PLS Project (Persano Adorno *et al.* 2017), two of our most relevant changes were related to: i) the discovery of interesting laboratory experiences of Modern Physics exploitable also at upper secondary school level (Petri & Niedderer 1998, Indelicato *et al.* 2013, Persano Adorno & Pizzolato 2015); ii) the knowledge of a new teaching methodology, the IBSE (Inquiry Based Science Education) approach (Herron 1971, Bybee 1993, McDermott 1996, National Research Council 2000, Llewellyn 2002, Banchi & Bell 2008), that can constitute a successfully teaching approach to effectively engage students into an active learning of the MP.

In this work we present the results of an inquiry-based teaching/learning path designed to overcome the difficulties of an exclusively theoretical approach to the introduction of Modern Physics (MP) topics. In particular, we planned and realized an inquiry-driven learning environment where about 20 students, from a second-year class of the vocational school "E. Fermi-F. Eredia" of Catania, were involved in the discovery of the Planck's constant, by following the steps indicated by Indelicato et al., 2013. The work, carried out as a cross-disciplinary module, has involved Physics and Technological Laboratory.

2. IBSE: why inquiry?

Inquiry-based Science Education (IBSE) – *learning through questioning* – is a teaching methodology that aims to promote the active participation of the students in the classroom through the use of scientific inquiry, towards a more direct understanding of "how" science effectively produces new knowledge. Inquiry is the "intentional process of diagnosing problems, critiquing experiments, and distinguishing alternatives, planning investigations, researching conjectures, searching for information, constructing models, debating with peers, and forming coherent arguments" (Byrnea & Johnstone 1987, NRC 2000, Llewellyn 2002, NRC 2012). Within an inquiry environment, the laboratory is not considered the place where students only observe experiences carried out by others or attend fruitless demonstrations of the validity of physics laws previously introduced by the teacher theoretically. The students are personally involved in experimental activities, facing problematic situations that requires reasoning efforts, in order to be solved effectively. Moreover, the laboratorial activity must be preceded by a preliminary phase characterized by posing scientifically relevant questions, designing procedures and be followed by a final critical evaluation of obtained results. Finally, scientific practices also include the sharing of ideas with peers, drawing explicatory models, supporting conclusions and making choices based on arguments and evidences.

The teaching strategies involved in inquiry approaches are grounded on the viewpoint that students are active thinkers, who construct their own understanding from interactions with phenomena, the environment, and other individuals. In inquiry-based learning, the students are engaged in identifying scientifically oriented questions, planning investigations, collecting data and evidences in laboratory and/or real life situations, building descriptions and explanation models, sharing their findings and eventually addressing new questions that arise. Depending on the amount of information and support provided by the teachers, the learners may be involved in a structured/guided inquiry or OI (Schwab 1962, Herron 1971, Banchi & Bell 2008). Generally, in structured inquiry the questions and procedures are provided by the teacher, and students generate their own explanations, supported by the evidence they have collected. In guided inquiry the teacher provides the students with only the research questions, and the students design the procedures to find reasonable answers and/or test the resulting explanations. In OI-based instruction, the teacher takes the delicate role of defining the context for inquiry, stimulating the students to derive their own questions, design and carry out independent investigations, construct coherent explanations, share their findings. This level of inquiry requires the highest capacity of scientific reasoning.

3. Activity description

The task assigned to the students was the design of an illuminated sign, efficient from the energetic point of view, flexible and attractive. The work, carried out as a cross-disciplinary module, has involved Physics and Technological Laboratory during 6 hour for week, lasting 8 weeks. Teachers acted as knowledge facilitators, providing support or materials, during all the teaching /learning process. The learning activity consisted of two different phases: the first one conducted the learners to the design and realization of the own signboard, causing them to reflect on the diode LED response; the second one, was focused on the calculation of the Plank's constant, mainly involving collection and analysis of the data. Each phase followed the 5E cycle (ENGAGE, EXPLORE, EXPLAIN, EXTEND, EVALUATE) in order to effectively develop student critical thinking and to help students to explore and evaluate their learning (Bybee 1993) (see Fig. 1).

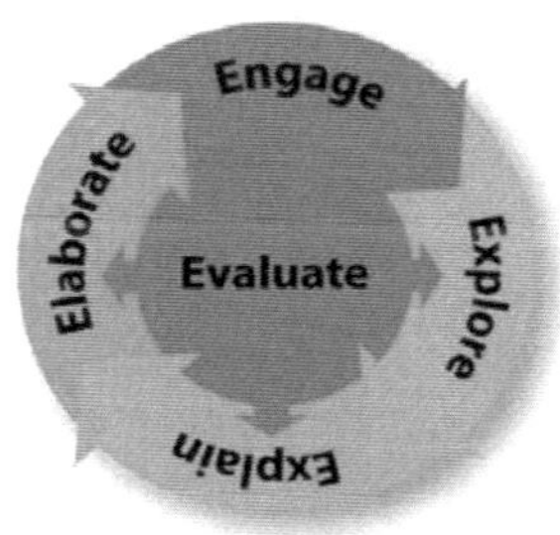

The 5E Model

Engage	Focuses student attention on the context and concepts of the unit.
Explore	Provides opportunities for students to examine their ideas through first hand experiences.
Explain	Introduces formal language, terms, symbols, and models.
Elaborate	Applies and extends students' developing ideas to new contexts.
Evaluate	Encourages students to assess their learning progress.

Fig. 1. The Inquiry 5E learning cycle

Phase 1

a. The engagement: the design of an illuminated sign

Students were asked to work in groups and to perform scientific investigations devoted to the design, realization and testing of illuminated signs, attractive and having physical characteristics able to maximize the efficiency. Through a *guided inquiry* environment, the students evaluated the different solutions, also by using internet resources to gather literature, developed their critical thinking and decided to implement the illuminated sign by using LED diodes.

b. Execution of experiments 1

Students were stimulated to carry out their own experimental work in the most independent way they were feeling confident to do it. During the first session in laboratory, our students discovered the laws governing the electrical connections. In particular, they learned how to connect the LEDs and determined the load resistance to avoid the LED "burning". Then each student designed its light signboard and carried out it. Students used the logbooks to note the followed procedure, the difficulties encountered throughout the activity and the changes they made during the inquiry process. Figs 2 and 3 refer to this phase.

Fig. 2. Students learn how connect LED

IOP Conf. Series: Journal of Physics: Conf. Series **1076** (2018) 012009 doi:10.1088/1742-6596/1076/1/012009

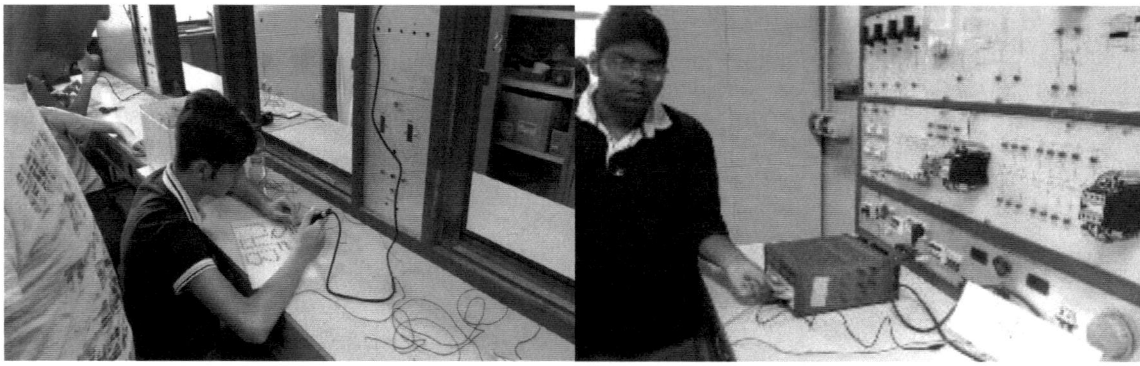

Fig. 3. Students realizing own light signboards

c. Cooperative work in the classroom (Explain, Elaborate, Evaluate)
During the cooperative work in the classroom the students commented the differences in the I-V characteristic curve between ohmic conductors and LED diodes, asking several interesting questions. In particular, they analyzed the collected data and built the I-V characteristic (see Fig. 4).

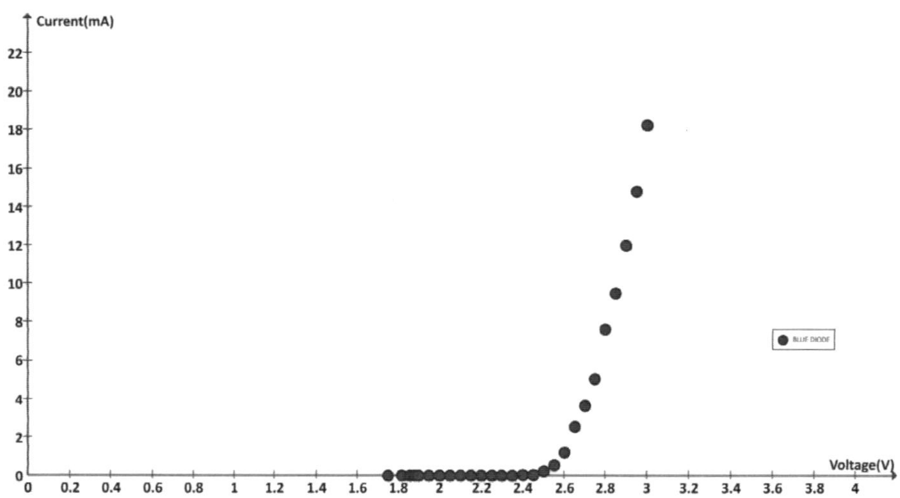

Fig. 4. The I-V curve in our Blue LED, $\lambda = 430$ nm

Phase 2
a. Frontal lesson (including video-projection, multimedia material, data sheets, discussion)
A deep reflection about the atomic models gave to the students the answers to the questions generated during the Phase 1. Moreover, the Bohr's model and the Einstein explanation of the photoelectric effect led the learners to the discovery of the Planck's constant. The approach with atomic model has allowed to discover the mechanism of the working of a LED diode They learned that a diode emits light when it reaches a threshold voltage and uncovered the "new" law

$$eV_s = hv$$

where e is electron charge, V_s is threshold voltage, h is Planck's constant and v is frequency. After this step the new challenging task of our students was the measurement of the Planck's constant.

IOP Conf. Series: Journal of Physics: Conf. Series **1076** (2018) 012009 doi:10.1088/1742-6596/1076/1/012009

b. Laboratorial activity 2

Fig. 5. Our experimental setup and students working in team

Students evaluated the different systems for the measurement of Planck's constant, making use of Internet resources gathered in literature. They planned how to make it during the technological laboratorial activities at the school. The students selected diodes with specific characteristics and purchased them. The experimental set-up was home-made by the students after they acquired critical thinking on it (see Fig. 5). In particular, they designed and carried out an apparatus formed by a set of six LEDs of different colors, with an external nominal load resistor of 100 Ω to prevent too high currents, a standard low-voltage power supply and two digital multimeters of standard quality. The electric circuit is shown in figure 6.

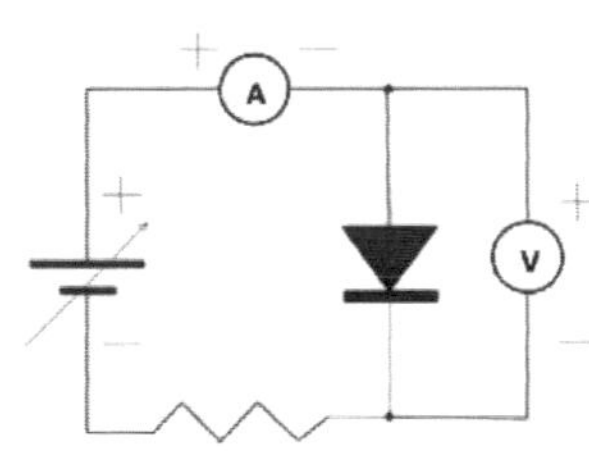

Fig. 6. The electric circuit of our experimental set-up

The students observed and registered the bias voltage applied to the LED and the relative currents values. The I-V curves carried out for the employed LEDs are shown in Fig. 7.

IOP Conf. Series: Journal of Physics: Conf. Series **1076** (2018) 012009 doi:10.1088/1742-6596/1076/1/012009

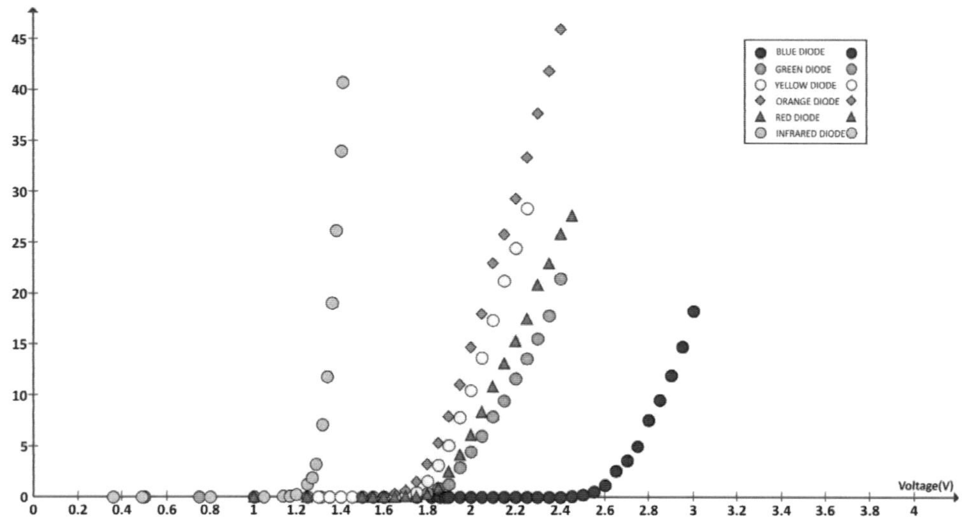

Fig. 7. I-V characteristic curves carried out for the employed LEDs

c. Cooperative data analysis and peer to peer work in the classroom

The analysis of data gathered from the I-V curves has raised several problems, mainly due to the fact that the threshold voltage was not well-defined. We cooperatively discussed on this problem and reflected on it. After long discussion, in order to overcome the problems of the identification of the "turn-ON" voltage, the students decided to follow two different criteria to determine it:

- to take the value of the bias voltage when the electrical current starts to be different from zero and LED was "turned ON" (*Method 1*);
- to extrapolate a small part of the nearly exponential curve linearly to zero (*Method 2*).

Analysis of the experimental findings

About a half of the students decided to use the first method for the determination of the of the "turn-ON" voltage and in table 1 we report the data collected by using this method.

Table 1. Threshold voltage, Energy, Planck's constant value f or the six LED (characterized by their wavelength λ and frequency ν) obtain by means of *Method 1*

Wave length λ **[nm]**	**Frequency** ν **[10^{14}Hz]**	**Threshold Voltage** V_s **[V]**	**Energy** **E [10^{-19}J]**	**Planck's Constant** **h [10^{-34} Js]**
430	6.98	2.50 ± 0.01	4.00	5.73
565	5.31	1.85 ± 0.01	3.51	6.61
590	5.08	1.70 ± 0.01	2.72	5.35
627	4.78	1.65 ± 0.01	2.64	5.52
700	4.29	1.73 ± 0.01	2.8	6.52
850	3.53	1.20 ± 0.01	1.87	5.29

Afterwards, the students calculated the average value of Planck's constant and its standard deviation Δh, obtaining **h = (5,84 ± 0,59)10^{-34} Js**.

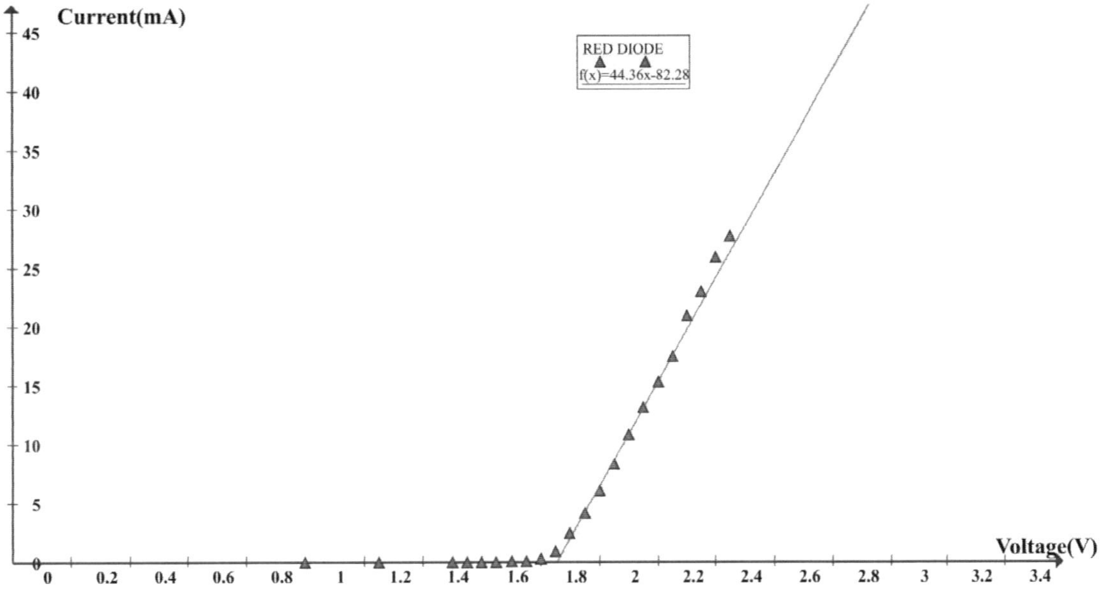

Fig. 8. Threshold voltage determined by means of a linear regression (*Method 2*)

The sample of students deciding to use the *Method 2*, carried out the best fit of the data in order to obtain the threshold voltage V_s (see Fig. 8). Afterwards, by means of a statistical analysis, they calculated the value of the Planck's constant for each LED. They also calculated the uncertainty for each value obtained. The values of Planck's constant together with their error bars are shown in Fig. 9. During this relevant phase of the cooperative work, the students learned the significance of linear regression of the experimental data and how to calculate the error attributable to each measure

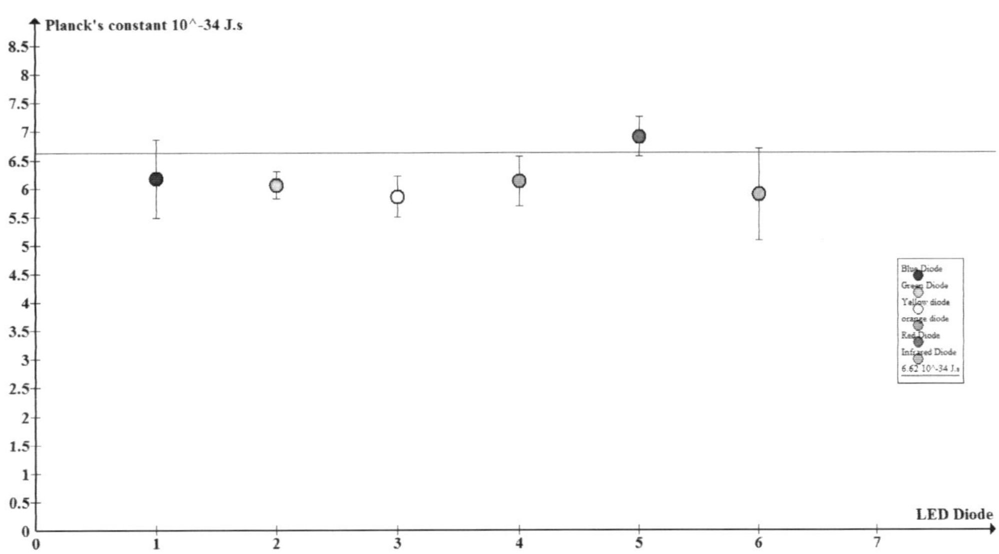

Fig. 9. Experimental values of Planck's constant together with their error bars, obtained with *Method 2*

In table 2 we report the data collected by using the *Method 2*

IOP Conf. Series: Journal of Physics: Conf. Series **1076** (2018) 012009 doi:10.1088/1742-6596/1076/1/012009

Table 2. Threshold voltage, Energy, Planck's constant value f or the six LED (characterized by their wavelength λ and frequency ν) obtain by means of *Method 2*

Wave length λ [nm]	**Frequency ν [10^{14} Hz]**	**Threshold Voltage Vs ± ΔV_s [V]**	**Energy E ± ΔE [10^{-19}J]**	**Planck's Constant h ± Δh [10^{-34} Js]**
430	6.98	2.69 ± 0.30	4.30 ± 0.48	6.17 ± 0.69
565	5.31	1.91 ± 0.08	3.06 ± 0.15	6.06 ± 0.24
590	5.08	1.86 ± 0.11	2.98 ± 1.81	5.85 ± 0.36
627	4.78	1.83 ± 0.13	2.93 ± 0.21	6.12 ± 0.44
700	4.29	1.85 ± 0.05	2.96 ± 0.15	6.91 ± 0.35
850	3.53	1.30 ± 0.18	2.08 ± 0.28	5.89 ± 0.81

Also in this case, the students calculated the average value of Planck's constant and its standard deviation Δh, obtaining **h = (6,16 ± 0,39)10^{-34} Js**.

At the end of the cooperative work, the students by means of peer to peer discussion in the classroom, compared the values for the Planck's constant obtained by using the two methods.

Both results are compatible with the accepted value of the Planck constant, but only that obtained with the second method stays within approximately the 10% of it, which is a reasonable result for an educational measurement.

4. Discussion and conclusion

Here we presented and discussed the outcomes of an inquiry-based teaching/learning path aimed at surmounting the difficulties of an exclusively theoretical approach to the introduction of MP topics in a vocational secondary school. Students learned how to conduct a scientific research activity, starting from an initial collection of information (literature), and moving across a planning phase, the design and realization of measurements, gathering and analyzing data, the formulation of hypothesis and modeling, drawing conclusions.

At the end of laboratorial activities we administered an open-ended questionnaire about basic MP arguments both to the students involved in this workshop (experimental group) and to a sample of students characterized by the same age and curricular instruction, but belonging to a different class not involved in this learning experience (control group). The comparison between the questionnaire marks obtained by the two student groups shows that the students which have attended the inquiry-based laboratorial learning path have achieved considerably higher scores both in terms of concept understanding and abilities in the synthesis, interpretation and evaluation of experimental data.

More important, at the end of this module the students acquired much more confidence in their approach to the study of physics, even under the point of view of their social skills, as a direct result of the teamwork carried out during the exploration phase. The presented laboratorial learning sequence has enhanced the student outcomes also in terms of motivation, interest, reflective participation, critical observation and discovery learning. Information about the student affective development and motivation to learn was achieved by means of structured interviews based on the Intrinsic Motivation Inventory, with specific items adapted to our study (Jang *et al.* 2016). Moreover, the learners have enthusiastically attended a summer Physics school held at the Scientific Lyceum "A. Volta" of Caltanissetta where they presented and shared with students of other high-schools the whole activity.

For the sake of saving space, we cannot report here the whole analysis of the student answers to both the questionnaire and satisfaction survey. The full analysis and results will be the subject of a forthcoming paper.

Fig. 10. A group of our students presented this laboratorial work to the students of different high-schools during a summer Physics school

In conclusion, the proposed inquiry-based activity provided students with the opportunity to develop a large range of complementary skills such as working in groups, synthesis, interpretation and evaluation of the experimental data, experience of open-ended problems solving and other cross-disciplinary abilities. Moreover this activity has had a positive impact on students' attainments, with an even stronger impact on the students with lower levels of self-confidence and those from disadvantaged backgrounds.

References

Banchi H. & Bell R. (2008) The many levels of inquiry. *Science and Children* **46**(2), 26-29.

Bybee R.W. (1993) An instructional model for science education. In *Developing Biological Literacy*, Biological Sciences Curriculum Study, Colorado Springs, CO.

Byrnea M.S. & Johnstone A.H. (1987) Critical Thinking and Science Education. *Studies in Higher Education* **12**(3), 325-339.

Etkina E. (2010) Pedagogical content knowledge and preparation of high school physics teachers. *Physical Review Special Topics Physics Education Research* **6**, 020110.

Foti A. & Gianino C. (1999) Elementi di analisi dei dati sperimentali-Liguori editori.

Francaviglia M., Lorenzi M.G., Michelini M. & Stefanel A. (2012) IDIFO3 – teachers formation on modern physics and mathematical foundations of quantum physics: a cross sectional approach. *Journal of Applied Mathematics* **5**, 231-240.

Global Science Forum (2008) *Encouraging student interest in science and technology studies*.

Herron M.D. (1971) The nature of scientific enquiry. *School Review* **79**, 171-212.

Indelicato V., Rocca P., Riggi F., Santagati G. & Zappalà G. (2013) Analysis of LED data for the measurement of Planck's constant in the undergraduate laboratory. *European Journal of Physics* **34**(4), 819-830.

Ireson G. (2000) The quantum understanding of pre-university physics students. *Physics Education* **35**, 15-21.

Jang H., Reeve J. & Halusic M. (2016) A New Autonomy-Supportive Way of Teaching That Increases Conceptual Learning: Teaching in Students' Preferred Ways. *The Journal of Experimental Education* **84**(4), 686-701.

Johnson I.D., Crawford K. & Fletcher P.R. (1998) Student difficulties in learning quantum mechanics. *International Journal of Science Education* **20**, 427-446.

Kroemer A. (1994) Investigating quantum physics in the high schools environments. *Physics Education* **56**, 127-139.

Llewellyn D. (2002) *Inquiry Within: Implementing Inquiry-based Science Standards*. Corwin Press Inc., Thousand Oaks, CA.

McDermott L.C. & Physics Education Group at the University of Washington (1996) *Physics by Inquiry*, John Wiley & Sons, New York.

Michelini M., Ragazzon R., Santi L. & Stefanel A. (2000) Proposal for quantum physics in secondary school. *Physics Education* **35**, 406-410.

Michelini M., Ragazzon R., Santi L. & Stefanel A. (2004a) Implementing a formative module on quantum physics for pre-service teacher training. In *Quality Development in the Teacher Education and Training* (Michelini M., ed.), Forum, Udine, pp. 429-435.

Michelini M., Ragazzon R., Santi L. & Stefanel A. (2004b) Discussion of a didactic proposal on quantum mechanics with secondary school students, *Il Nuovo Cimento* C **27**, 555-567.

Michelini M. (2010) *Progetto IDIFO. Fisica Moderna per la Scuola. Materiali, aspetti e proposte per l'innovazione didattica e l'orientamento*, MIUR-PLS-UniUD, Udine.

Muller R. & Wiesner H. (2001) Teaching quantum mechanics on an introductory level, *American Journal of Physics* **70**, 200-209.

National Research Council (NRC) (2000) *Inquiry and the National Science Education Standards: A Guide for Teaching and Learning*, The National Academies Press, Washington DC.

National Research Council (NRC) (2012) *A Framework for K-12 Science Education: Practices, Crosscutting Concepts, and Core Ideas*, The National Academies Press, Washington, DC.

Olsen R.V. (2002) Introducing quantum mechanics in the upper secondary school: a study in Norway, *International Journal of Science Education* **24**, 565-574.

Petri P. & Niedderer H. (1998) A learning pathway in high-school level quantum atomic physics, *International Journal of Science Education* **20**, 329-347.

Persano Adorno D. & Pizzolato N. (2015) An inquiry-based approach to the Franck-Hertz experiment. *Il Nuovo Cimento* C **38**, 109. DOI: 10.1393/ncc/i2015-15109-y.

Persano Adorno D., Fazio C., Pizzolato N. & Battaglia O.R. (2017) Training Pre-service and In-service Secondary School Teachers: Analysis of Changes in Perceptions About Quantum Physics Concepts and NoS Views. In *Key Competences in Physics Teaching and Learning* **190**, Springer Proceedings in Physics, pp. 165-176.

Pospiech G. (2000) A modern course in quantum physics for teacher education. In *Turning the challenge into opportunities* (Xingkai L. & Kaihua Z., eds.), Guangxi Normal University Press, Guilin, China, pp. 244-248.

Schwab J.J. (1962) The teaching of science as inquiry. In *The teaching of science*, (Schwab J.J. & Brandwein P.F., eds.), Harvard University Press, Cambridge, MA, p. 3-103.

IOP Conf. Series: Journal of Physics: Conf. Series **1076** (2018) 012010 doi:10.1088/1742-6596/1076/1/012010

Discussing fundamental topics of quantum physics using visualizations of bound states

Tommaso Rosi, Stefano Oss, Pasquale Onorato

University of Trento

Abstract. Many authors outlined the students' difficulties in the conceptual understanding of quantum physics. One of the major problems is that students find the mathematical language used to describe these systems (that cannot be seen) generally too obscure. A lot of research has been done to use computer visualizations of quantum mechanical topics and it proved to be a powerful and flexible tool, allowing to see through the mathematics.

In this work it is shown how the use of different visualizations regarding bound states (and in particular the hydrogen atom) may be useful to foster a deeper understanding of topics of very general interest for the study of quantum physics.

1. Introduction

There are many reasons why quantum physics is such a difficult subject both to understand and teach. It is a very abstract subject, and the mathematics used by scientists to describe quantum systems is quite advanced and certainly not accessible at high school level or by a general public's audience (Bao & Redish 2002). The principles and possible interpretations of quantum physics also often oppose the common perception of the world, based on abstractions such as locality and causality. As a matter of fact, it is impossible for students (as for any human) to have any direct experience of the quantum world, and it is difficult – and not without complications – even to refer to pictures, videos or experiments to fill this gap.

Computer visualizations proved to be a flexible and powerful cognitive tool for teaching quantum physics (Catalohlu & Robinett 2002, Dwyer 1972, Gilbert 2005, McCormick *et al.* 1987, Robinett, 2000), as it directly translates complex mathematical equations describing systems that cannot be seen into interactive, visual representations. Teachers may then focus on discussing the physics and the general principles underlying those systems using intuitive and fascinating images to help both developing and memorizing mental models about the discussed topics (Johnston *et al.* 1998, Karplus 1977).

It is very important to be able to develop educational approaches to quantum physics that are based on hands-on activities rather than on a more traditional kind of lecture in which students are simply sitting and passively listening (McKagan *et al.* 2008, Zollman *et al.* 2002), so interactivity play a major role in the design of educational visualizations and simulations. User interfaces and visual representations of the physics need to be research based (McKagan *et al.* 2008) to increase the effectiveness of the activities and, most of all, to avoid misinterpretations and misconceptions that may arise from a bad design. Education research and direct teaching experience help understanding the students' main difficulties and misconceptions, so that they can be the targeted by computer based activities (Kohnle 2012, Kohnle *et al.* 2014). Research based on the sum over paths approach and Geo Gebra simulations has also been made (Malgeri *et al.* 2014).

In this work we focus on ways of presenting topics of very general interest about quantum physics based on visualizations of bound states, and especially on the hydrogen atom (Rosi & Oss 2015). In

IOP Conf. Series: Journal of Physics: Conf. Series **1076** (2018) 012010 doi:10.1088/1742-6596/1076/1/012010

fact, the hydrogen atom is a very interesting system to focus on. First of all, it is a very important system for historical reasons and it is mentioned at any level of QM didactics. Most importantly, it is a real system for which simple low-cost experiments may be done (Onorato *et al.* 2015) (differently from square wells or the harmonic oscillator).

2. Fundamental topics that can be discussed

What is a wavefunction? Where is the electron?

First of all, the concept of wavefunction is one of the most important and fundamental ones, as it is a crucial ingredient in connecting theory and experiments, to be able to discuss about measurements, probability, uncertainty and so on (Bao & Redish 2002, Chhabra & Das 2016, Singh 2001). Visualizing a wave function may be a good way to start discussing what it represents: there is no need of using formal mathematical tools and it is possible to focus on the meaning of a probability density.

1D infinite and finite wells are generally discussed and many visualizations regarding them are available (even for the 2D case) (Thaller; Falstad; "PhET"; "QuVis"; "Physlet Quantum Physics"). For example, "Quantum Bound States" by "PhET" allow users to "explore the properties of quantum 'particles' bound in potential wells" letting you see "how the wave functions and probability densities that describe them evolve (or don't evolve) over time".

A very interesting system, the one that we prefer and that we suggest to start with, is the hydrogen atom (Rosi & Oss 2015, Thaller, Falstad, "Hydrogen!", "Atom in a Box"). Students probably already have a mental representation of it – an electron orbiting around a proton – so teachers need to focus on the difference between orbits and orbitals. To help teachers with this, the hydrogen atom orbitals may be seen in classroom, and students can interactively choose which one to see and look at them from different perspectives. Some simple features of the hydrogen atom may easily be seen: for example, it can be noticed simply by changing the quantum numbers that the size of orbitals gets larger as the corresponding energy gets larger too, as shown in Fig. 1.

Presenting hydrogen orbitals may also come in handy when discussing spectroscopy, as it will be seen later.

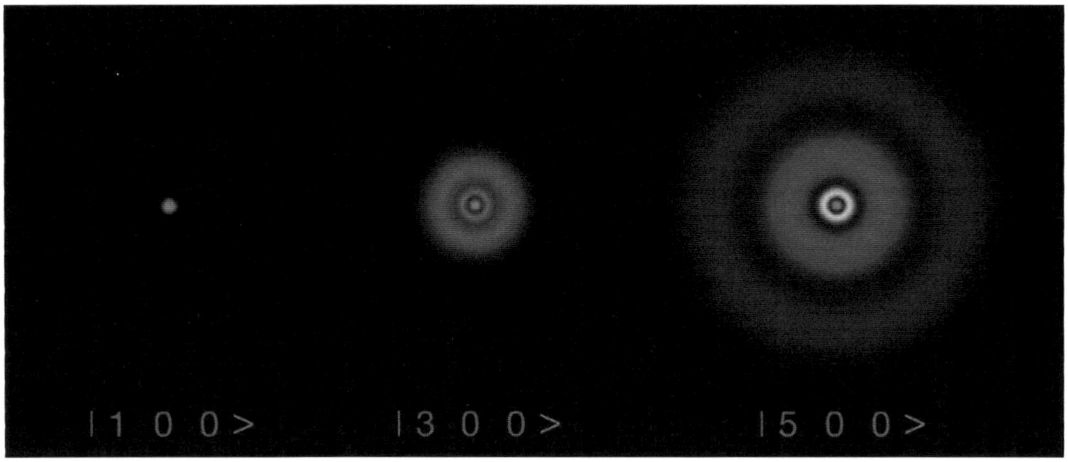

Fig. 1. Examples of hydrogen orbitals (Rosi & Oss, 2015; "Hydrogen!")

The correspondence principle

It is very important to be able to connect the quantum world to the classical results students have being studying all along.

Other authors spent much efforts to show how under the right conditions quantum objects will behave as their classical counterparts e.g. using the sum over paths method: applying this method in the

GeoGebra environment students are able to get results of classical mechanics and geometrical optics working with quantum objects (Malgeri *et al.*, 2014).

It is possible to refer to visualizations of the hydrogen atom again to build a bridge between classic and quantum physics, meaning to get an intuitive idea of the Correspondence Principle: in fact, it can be seen that a hydrogen orbital with large quantum numbers resembles a classical orbit (Rosi & Oss 2015, "Hydrogen!", Falstad, "Atom in a Box"), as shown in Fig. 2. Of course it is fundamental to specify the differences between the two models, as in the quantum mechanical orbital the electron is still delocalized over it, a highly non-classical behaviour.

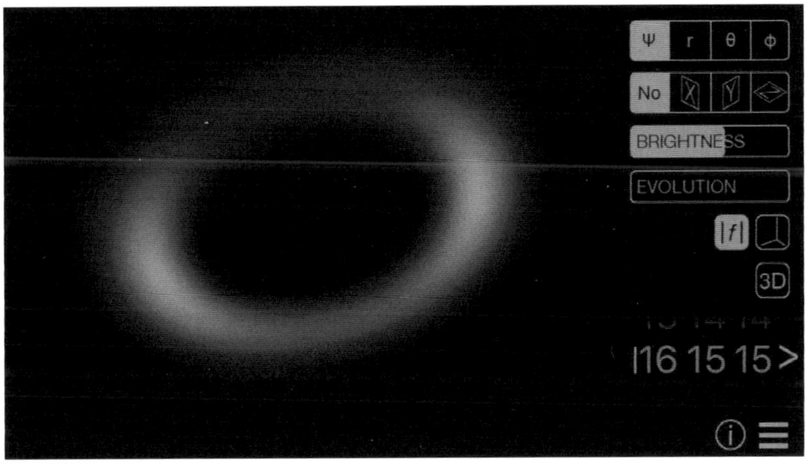

Fig. 2. For large quantum numbers a hydrogen orbital resembles a classical orbit. It is here shown the modulus of the n = 16, l = 15, m = 15 orbital (Rosi & Oss 2015, "Hydrogen!")

Superposition and interference

Superposition and interference play major roles in quantum physics. They are the key concepts underneath the double slit experiment which is probably the most cited experiment in introductory quantum mechanics courses. These concepts may as well be discussed using interactive visualizations that allow teachers and students to create superpositions of wave functions and see how they interfere with each other, meaning to visualize the "complex" nature of quantum wave functions. This allows to focus on the difference between classical and quantum probability, analogously to what is done presenting the double slit experiment.

It is possible to create superpositions of states using many different applets (see for example "Superposition states in an infinite square well" from "QuVis" or "Quantum Bound States" from "PhET").

Again, this interference of states may also be seen in the hydrogen atom (Rosi & Oss 2015, "Hydrogen!", Falstad). See for example Fig. 3: here the sum of two orbitals, namely the n = 3, l = 2, m = –2 and the n = 3, l = 2, m = 2 is shown on their right. Interestingly, in some regions the summed orbitals cancel out and the result goes to zero (destructive interference) while in others the result is larger than the starting orbitals (constructive interference). A more advanced subject may also be the representation of a particle in a harmonic potential using superpositions of the 2D harmonic oscillator eigenfunctions (Falstad).

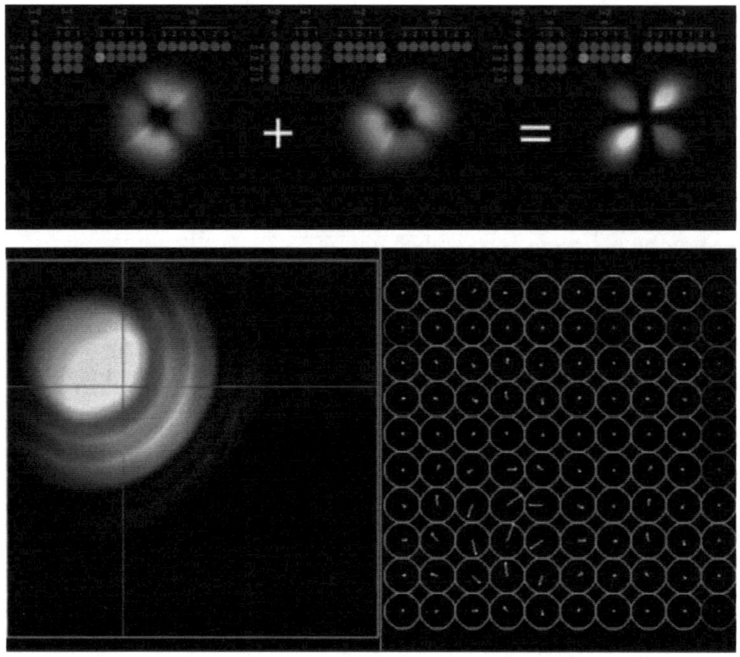

Fig. 3. Examples of visualisations of superpositions. (Top) Interference between two hydrogen orbitals (Rosi & Oss 2015, "Hydrogen!"). (Bottom) Simulation of a particle in a two dimensional rectangular square well (Falstad)

Stationarity and non-stationarity

The time evolution of such superpositions (and their moduli) may be a challenge to imagine, but they certainly may be visualized. Choosing appropriate superpositions it is straightforward to visualize the conceptual meaning of the difference between a stationary and a non-stationary state. For example, in Fig. 4 it is shown the time evolution of the modulus of a linear combination of three hydrogen orbitals with different energies: since the linear combination results in a non-stationary state (the rate of the rotation of the phase of the orbitals is proportional to their energy), its shape changes in time. The shape of a stationary state (for example of a single orbital) instead, never changes in time.

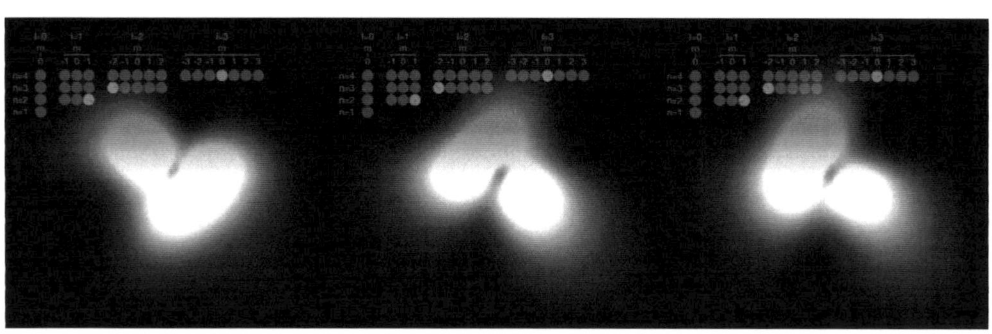

Fig. 4. The modulus of a superposition of states with different energies may be visualised to see the evolution of a non-stationary state (Rosi & Oss 2015, "Hydrogen!")

Permitted transitions and spectroscopy

Energy quantization and energy level diagrams can be confusing for students, and it is often counter-intuitive for them to accept that energies of photons coming from a lamp depend on the energy difference between two levels (Zollman *et al.* 2002). It can be really rewarding to do some spectroscopy experiments and then use software visualizations to help them understand how energy levels models can correctly explain their observations ("VQM"; "Hydrogen!"; "PhET"; "Physlet Quantum Physics", "Atom in a Box"). For example, the "Emission Spectroscopy" applet from "VMQ" (shown in Fig. 5, on the left) allows students to look at spectra of different gas lamps and to interactively adjust energy levels to try to get the correct transitions by trial and error. On the right of the same figure, instead, the "Dipole Transitions" mode of "Hydrogen!" is shown, in which for a chosen starting orbital all transitions allowed by selection rules are displayed. With this mode it is possible to see, for example, that the 2s state cannot have any orbital with lower energy for which a dipole transition exists: this is why it is a so-called "metastable state", meaning that it has a longer lifetime than the other excited orbitals.

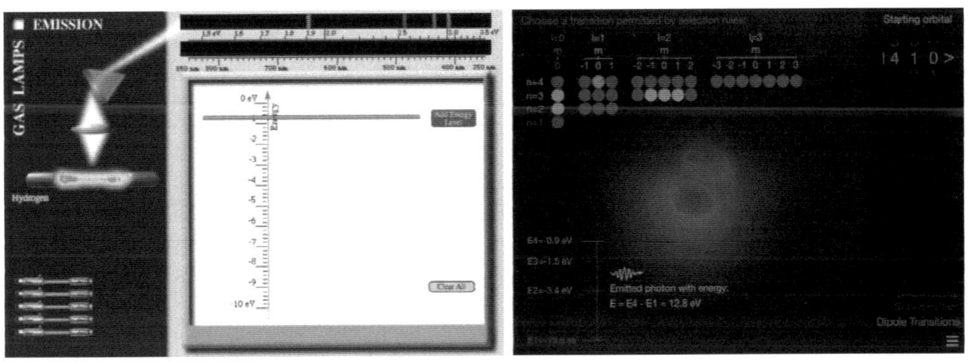

Fig. 5. Visualisations may be useful to understand theoretical models based on energy levels, especially when used after some spectroscopy in the laboratory has been done. (Left) "Emission Spectroscopy" applet from "VMQ" (Zollman *et al.* 2002). (Right) "Dipole Transitions" mode of "Hydrogen!" (Rosi & Oss 2015, "Hydrogen!")

3. Conclusions

We showed how the use of visualisations regarding bound states may be useful to discuss many topics of very general interest for the study of quantum physics. Particular attention was paid for visualisations of the hydrogen atom, a very important system of which some simple, low-cost experiments may be done in classroom.

First of all, visualisations are helpful in introducing the concept of wave functions, helping to focus on their link with probability density. For example, using the hydrogen atom, the teacher may focus on the difference between classical orbits and orbitals, and thus the importance of reasoning in terms of probability when dealing with quantum objects.

The capability of interactively creating superpositions of states is really helpful as students may see the influence of interference on such superpositions, analogously to what happens in the double slit experiment. Visualizing the time evolution of superpositions also allow to get an intuitive idea of the difference between a stationary and a non-stationary state.

An intuitive idea of the Correspondence Principle may also be given looking at a single hydrogen orbital with large quantum numbers, as it becomes more and more similar to a classic circular orbit as its corresponding quantum numbers get larger and larger (even though particular care must be put in emphasising the differences between the two models that still exist).

As a last example, the so-called "permitted transitions" of the hydrogen atom may be displayed, allowing to take a further step in the exploration of the discrete–dominated world of bound states. It was also pointed out that as this subject is discussed, it can be very rewarding to also do some spectroscopy in the laboratory, so that a bridge between experimental observations and theoretical models can be built, as such bridges can play a very important role in the learning process.

IOP Conf. Series: Journal of Physics: Conf. Series **1076** (2018) 012010 doi:10.1088/1742-6596/1076/1/012010

References

Atom in a Box. http://daugerresearch.com/orbitals/index.shtml.

Bao L. & Redish E.F. (2002) Understanding probabilistic interpretations of physical systems: a prerequisite to learning quantum mechanics, *American Journal of Physics* **70**, 210-217.

Cataloglu E. & Robinett R.W. (2002) Testing the development of student conceptual and visualization understanding in quantum mechanics through the undergraduate career, *American Journal of Physics* **70**, 238-251

Chhabra M. & Das R. (2016) Quantum mechanical wave function: visualization at undergraduate level, *European Journal of Physics* **38**, 015404-015416.

Dwyer F.M. (1972) The effect of over responses in improving visually programming science instruction, *Journal of Research in Science Teaching* **9**, 47-55.

Falstad P., Quantum mechanics applets. http://www.falstad.com/mathphysics.html.

Gilbert J.K. (2005) Visualization in Science Education. Amsterdam: Springer. DOI: 10.1007/1-4020-3613-2.

Johnston I.D., Crawford K. & Fletcher P.R. (1998) Student difficulties in learning quantum mechanics, *International Journal of Science Education* **20**, 427-446.

Karplus R. (1977) Science teaching and the development of reasoning, *Journal of Research in Science Teaching* **14**, 169-175.

Kohnle A., Baily C., Hooley C. & Torrance B. (2014) Optimization of Simulations and Activities for a New Introductory Quantum Mechanics Curriculum, PERC Proceedings, 209-212. DOI: 10.1119/perc.2013.pr.040.

Kohnle A., Cassettari D., Edwards T.J., Ferguson C., Gillies A.D., Hooley C.A., Korolkova N., Llama J. & Sinclair B.D. (2012) A new multimedia resource for teaching quantum mechanics concepts, *American Journal of Physics* **80**. DOI: 10.1119/1.3657800.

Malgieri M., Onorato P. & De Ambrosis A. (2014) Teaching quantum physics by the sum over paths approach and GeoGebra simulations, *European Journal of Physics* **35**, 055024-055044.

McCormick B.H., DeFanti T.A. & Brown M.D. (1987) Visualization in scientific computing, *Computers & Graphics* **21**, 14-18.

McKagan S.S., Perkins K.K., Dubson M., Malley C., Reid S., LeMaster R. & Wieman C.E. (2008) Developing and researching PhET simulations for teaching quantum mechanics, *American Journal of Physics* **76**, 406-417. DOI: 10.1119/1.2885199.

Onorato P., Malgieri M. & De Ambrosis A. (2015) A Measuring the hydrogen Balmer series and the Rydberg's constant with a homemade spectrophotometer, *European Journal of Physics* **36**, 058001.

QuVis. http://www.st-andrews.ac.uk/~www_pa/quvis/flash.html.

PhET interactive simulations. https://phet.colorado.edu/.

Physlet Quantum Physics – An Interactive Introduction. http://www.compadre.org/pqp/.

Robinett R.W. (2000) Visualizing the collapse and revival of wave packets in the infinite square well using expectation values, *American Journal of Physics* **68**, 410-420.

Rosi T. & Oss S. (2015) A bit of Quantum Mechanics, *The Physics Teacher* **53**, 230-233.

Rosi T., Hydrogen! iOS app. https://augmenteddidactics.wordpress.com/hydrogen/.

Singh C. (2001) Student understanding of quantum mechanics, *American Journal of Physics* 69, 885-895.

Thaller, B., Visual Quantum Mechanics. http://vqm.uni-graz.at/index.html.

VQM project by the Physics Education Research Group at Kansas State University. http://phys.educ.ksu.edu/Research/.

Zollman D.A., Rebello N.S. & Hogg K. (2002) Quantum mechanics for everyone: Hands-on activities integrated with technology, *American Journal of Physics* **70**, 252-259. DOI: 10.1119/1.1435347.

IOP Conf. Series: Journal of Physics: Conf. Series **1076** (2018) 012011 doi:10.1088/1742-6596/1076/1/012011

From one slit to diffraction grating: optical physics lab by means of computer on-line sensors

Daniele Buongiorno, Marisa Michelini, Lorenzo Santi, Alberto Stefanel

Università degli Studi di Udine

Abstract. Diffraction is a crucial phenomenon and its interpretation bridges from geometrical to wave optics and from wave optics to quantum mechanics. Becoming familiar with the characteristics of the diffraction in the cases of one, two, many slits is an important experience for students not only from the subject point of view, but also on the methodological plan. The exploration of the relative interpretation by means of simulation and modelling offers to the students the opportunity to experience the typical methodological work in physics. An educational proposal was developed for the study of optical diffraction: from the analysis of a single slit diffraction, to a double slit and to a diffraction grating. It is based on a USB acquisition system designed and developed for experimental data acquisition in an educational lab, correlating position and light intensity measurements in one direction (Gervasio & Michelini 2009). Data can be exported in text format and data fitting can be done by means of a spreadsheet. The educational proposal about single-slit diffraction (Corni *et al.* 1993) has been tested with 114 students (aged 18-19) in 2008 and 2014. Many of these interventions were research-based educational path. Data concerning learning processes of 29 students are described in (Michelini & Stefanel 2015). We are now developing an educational path about optical spectroscopy for upper secondary-school and university students, and the need to extend the educational proposal to the case of diffraction from a diffraction grating has emerged: this is necessary in order to deal with lab activities involving the analysis of spectra and to study the conceptual knots emerged in the literature in the case of optical spectra generated by a prism (Ivanjek *et al.* 2015a, 2015b). The phenomenon laws obtained by data are interpreted under the wave nature of light by students and these laws are used for spectroscopic analysis of different light sources. Simulation software allows to build models of interpretation of phenomenology based on the first principles (Santi *et al.* 1993). The proposed IBL path based on the exploration of the multiple slit diffraction pattern has been used with 98 upper secondary school Italian students and 56 freshmen in biotechnology sampled from University of Udine. Data has been analyzed and the results are under discussion. This poster shows the experimental activity proposed to the students and its characteristics.

1. Motivations and basis of the proposal

The previous proposal: diffraction of light from a single slit

The single-slit diffraction educational proposal aims to obtain the formal laws describing the phenomenon by means of a gradual phenomenological analysis of the data acquired via on-line measurement with the Lucegrafo system (Fig. 1): this USB system for diffraction measurements is constituted by a photodiode inserted in a housing solid with the cursor of a linear potentiometer, so that the optic signal is correlated with the position by means of the resistance of the potentiometer (Fig. 1 left). The system acquires and represents on the screen, both in graphical and numerical way, couples intensity-position so that, moving the cursor by manual adjustment of the screw, the space distribution of light intensity on the screen is acquired (Fig. 1 right). The intensity in the graph is represented in arbitrary units, proportional to the light intensity incident on the sensor. Data can be exported in text format and further analysis can be done by means of a spreadsheet.

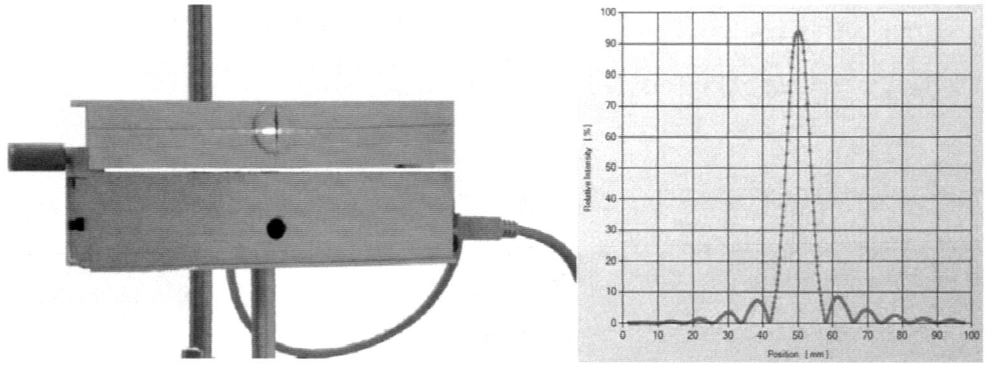

Fig. 1. The Lucegrafo system and the acquired graph light intensity vs position. Single slit (width a = 0,12 mm) at a distance of 155 cm from the screen with a laser (λ = 650 nm)

Students are involved in an experimental problem solving activity and then in an interpretative challenge by means of the following steps:

1. Qualitative analysis of the diffraction pattern allows them to recognize that it is an angular distribution, constituted by a sequence of maxima and minima of intensity symmetrically distributed with respect to the position of the central maximum.
2. Acquisition of the light intensity vs the transverse position using the on-line system Lucegrafo.
3. Analysis of the relation between angular position of minima and maxima as a function of their order with the help of a spreadsheet (Fig. 2).

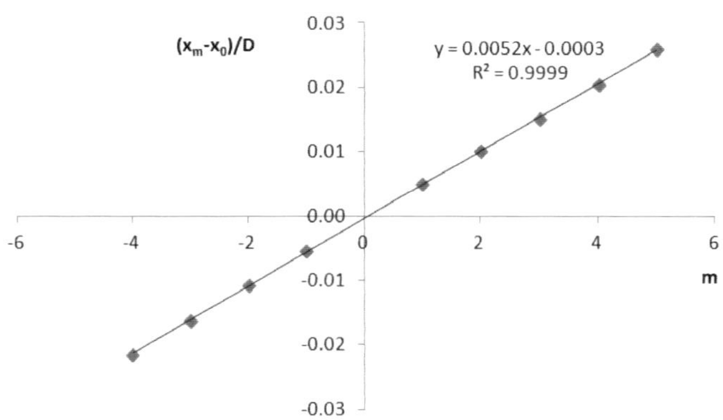

Fig. 2. Angular positions of minima vs order m. D is the screen-slit distance, X_0 is the position of the central maximum and x_m is the position of each minimum. The relation is linear

Phenomenological exploration (step 1) and data analysis from the on-line acquisition (steps 2-3) lead to discover that the slit width a is a fundamental parameter. From the phenomenological laws (Fig. 3) another fundamental parameter emerges, and it must have the dimension of a length and his a physical quantity characterizing the light used (usually indicated with λ).

IOP Conf. Series: Journal of Physics: Conf. Series **1076** (2018) 012011 doi:10.1088/1742-6596/1076/1/012011

$$\left.\begin{array}{l}\sin\theta = m\dfrac{\lambda}{a}\\ \sin\theta \approx \dfrac{x_m - x_0}{D}\end{array}\right\} \Rightarrow \frac{x_m - x_0}{D} \approx m\frac{\lambda}{a} \quad m = \pm 1, \pm 2, \ldots$$

Fig. 3. The slope in the graph of Fig. 2 depends both on the slit width a and on another parameter named λ. This is associated with the color of the light. In an wave model it is the wavelength

4. Analysis of the relation between intensity of the maxima and their position (Fig. 4), as bridge to interpretation.

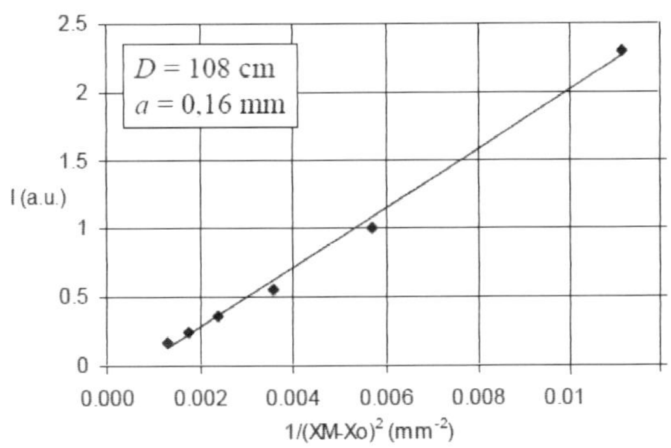

Fig. 4. Intensity of maxima vs their position (x_M-X_0). The relation is linear if one plots the intensity as a function of the squared inverse position

5. Reconstruction of the diffraction pattern trough models/simulations based on the Huygens-Fresnel and the superimposition principles (Fig. 8). Applying these principles accounts for the distribution observed, helping the interpretative process.

Interpretative problems about single slit diffraction

Following the single slit diffraction proposal sketched before, research-based experiments in the schools were carried out with 114 Italian high school students using tutorial monitoring the learning process. The analysis of data concerning learning processes (Michelini & Stefanel 2015) showed that more than 80% of students become familiar with the new phenomenology and the phenomenological relations describing it. However, the relationship between experimental results and the interpretative model based on the Huygens-Fresnel principle remained an open problem: about 60% of students used the phenomenological laws to remain on a descriptive plan; about 30% attempted to explain the features of the phenomenon using a geometrical model, focusing only on those aspects that can be justified within that model (i.e. the "enlargement of the beam"); 10% evidenced that a new model is needed.

Widening the context to gain interpretative ownership: from a single slit to diffraction grating

As pointed out in literature (Vosniadou 2008), the context is crucial in gaining the conceptual understanding in the learning process. In particular, it activates the interpretative elements. Assuming that build the basis for interpretative inquiry is of fundamental importance, we hypothesize that the application of the wave model in a wider and more complex contexts, faced once again on a phenomenological plan, can activate the students' gain of a new interpretative perspective. This imply to study diffraction using two, three slits and a diffracting grating, to analyze the role of the various parameters, as the number of slits or their separation in order to gain descriptive and interpretative ownership of the phenomena concerning a diffraction grating. In this perspective we developed a new proposal,

IOP Conf. Series: Journal of Physics: Conf. Series **1076** (2018) 012011 doi:10.1088/1742-6596/1076/1/012011

described in the following. We start from our researches, that evidence the difficulties of students in changing interpretative model, and from that of McDermott's group, that showed the need to specify the role of the diffraction grating (Ivanjek *et al.* 2015a, 2015b). Our research questions are:

- RQ1: How does experimental activity contribute in identifying the new phenomenology and the model underlying it?
- RQ2: How does experimental activity contribute in identifying the role of the various parameters?
- RQ3: How does experimental activity contribute in identifying the role of a diffraction grating?

2. The experimental activity

The interpretative problems emerged in the single-slit experimentations are mainly related to the Huygens-Fresnel principle. To produce a personal involvement and a gradual understanding we suggest an experimental-based activity in which students gradually face optical phenomena whose interpretation require the understanding of the wave superimposition principle. In particular increasing the number of slits (steps 1-4 below) allows further investigations, related to the resolving power of a diffraction grating (step 5). The activity lab work and simulation activities interplay in order to offer a gradual understanding, both on the phenomenological and interpretative plan, of the main characteristics of the diffraction pattern caused by the interaction of light with different obstacles.

Light diffraction from 2 and 3 slits

1. The first step is linked to the previous path replacing the single slit with a double slit. The pattern is the convolution of a pure interference pattern, modulated by a diffraction sync, easily detectable with the Lucegrafo system (Fig. 5). The interference pattern shows new regularities, as the distance between two maxima, that can be studied in function of new parameters: the number of slits and their spatial density.
2. The distance between the two slits has consequences on the resulting distance between contiguous maxima of the interference pattern, in particular, reducing the distance leads to a greater linear separation between two continuous maxima.
3. An increasing number of slits (three rather than two) shows that, all conditions being equal, the maxima remain in the same positions but they become sharper and more intense (Fig. 6).

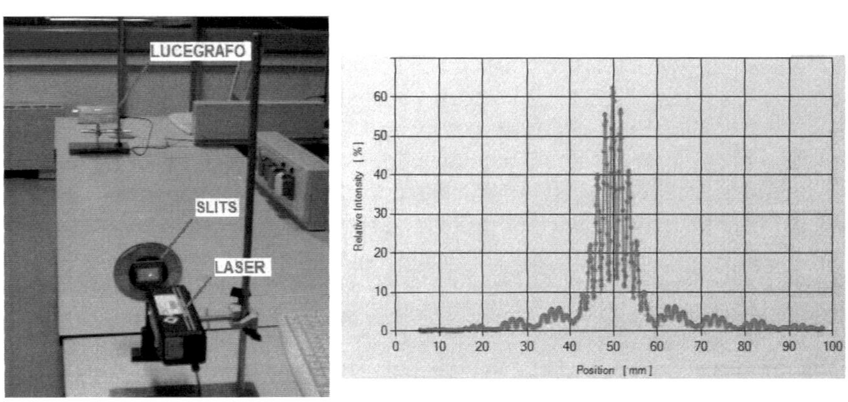

Fig. 5. Experimental setup and non-zero-width double slit diffraction pattern as obtained with the Lucegrafo system

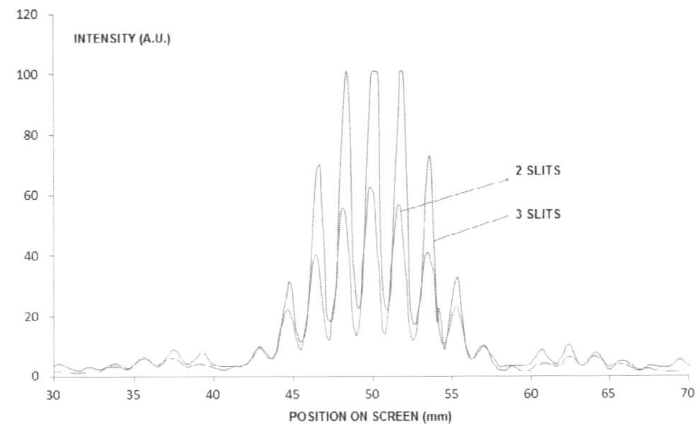

Fig. 6. Comparison between two and three slits patterns. Data obtained with the Lucegrafo system, processed with a spreadsheet (Excel)

4. Those observations, together with the fact that the intensity pattern depends on the color of the light, outlines that a large number of little-spaced slits allows to distinguish two maxima of different wavelengths of the same order.
5. The previous explorations provide the phenomenological basis to guess the characteristics of a diffraction pattern caused by a diffraction grating, characterized by a high density of narrow slits. The maxima are very sharp and highly-spaced, and the features of the interference pattern prevail on the diffraction ones (Fig. 7). If a superposition of wavelengths is used, the grating allows therefore effective maxima separation, and the possibility to carry out spectroscopic analysis.

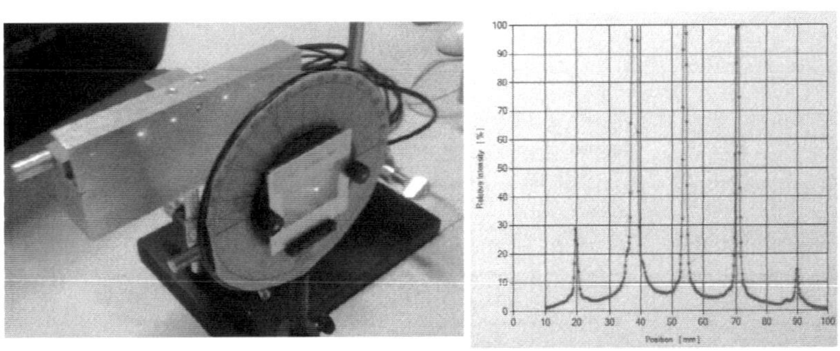

Fig. 7. Diffraction of laser light (λ = 650 nm) from a grating (300 lines/mm). Acquisition with the Lucegrafo system

Modeling and simulation

Beyond phenomenological exploration, modeling and simulation provide effective tools to support the proposal.

In modeling, a physical situation is represented by means of an abstract system described by the equation relating the variables, starting from first principles. In this perspective, a C++ script has been written (Santi *et al.* 1993): given the number of slits, their width and separation, the distance between them and the screen and the number of secondary sources along every slit, the software computes the superposition of the waves on every point of the screen using a wave hypothesis on the nature of light and the Huygens principle (Fig. 8 upper left). The same results can be obtained using a spreadsheet (Fig. 8 upper right).

On the other hand, in a simulation environment, students do not have access to the model, but they can appreciate in real time the consequences of a change in the values of the parameters (wavelength of the light, number, distance and width of slits) and they can see how a diffraction pattern changes changing the parameters. An example of such a simulation (Fig. 8 bottom) has been developed using Easy Java Simulation.

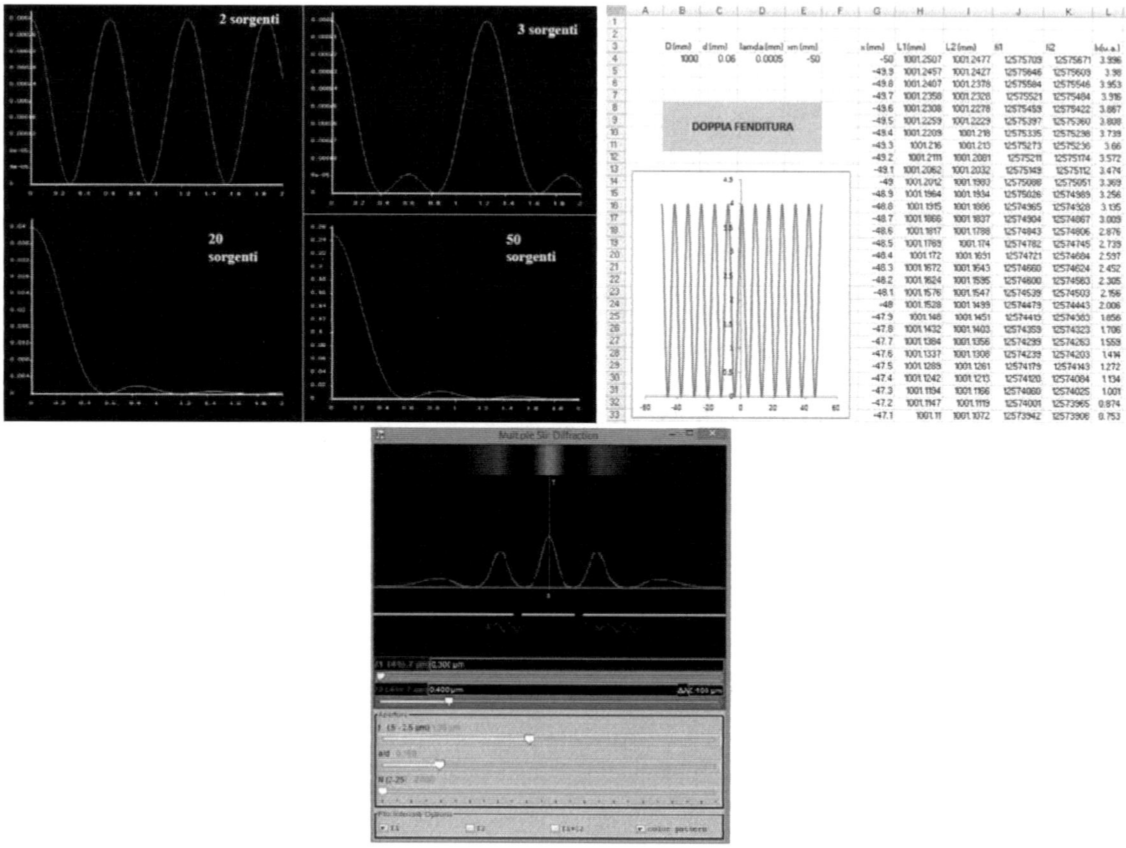

Fig. 8. Modeling: output of the C++ script (upper left) and calculus of a double-sit interference pattern by means of a spreadsheet (upper right). Simulation (bottom): user panel of the Easy Java Simulation (http://www.compadre.org/osp/items/detail.cfm?ID=8888). Parameters that can be adjusted are the number of slits, their width and their separation

3. A case study

Case study on the lab activity

The proposed path has been used as an interactive lecture demonstration with 98 upper-secondary school Italian students and 56 freshmen in biotechnology sampled from University of Udine (IT). In particular in the 2016 Summer School in Modern Physics at University of Udine it has been used as an interactive lecture demonstration after the experimental activity about single-slit diffraction proposed to 32 selected talented upper-secondary school students. Before the activity, a pre-test was submitted to all students. Concerning the present contribution we consider a question asking to say what had to be changed in an experimental setup in order to see a line spectrum rather than a continuous one. Moreover, here we discuss a case study carried out by means of detailed analysis of the final reports on experimental activity, written by 4 of them. The experimental activity has been proposed in a conceptual form, together with a paper for data collection, rather than a laboratory sheet. Therefore students wrote an argumentative synthesis outlining the conceptual framework of the experiment completing it with graphs and tables of data acquired.

Main learning outcomes

In a multiple-choice pre-test question students were asked to say what had to be changed in an experimental setup in order to see a line spectrum rather than a continuous one. Students evidenced difficulties in recognizing the role of the various elements: 14/32 students would not change the light source in order to see a line spectrum, but they would change the grating (5), the width of the slit (4) or the distance from the screen (5). In the reports students focused mainly on:

- Describing the pattern, that is: a symmetrical (4/4) angular distribution (2) of dark and bright fringes (3) in which the central maximum is the brightest (3).
- Mentioning the wave nature of light (2) and the Huygens-Fresnel's principle (4) to account for the observed pattern in terms of secondary sources (3) and wave fronts as envelopes (2). The principle is qualitatively described. A quantitative description that accounts for the observed pattern is missing.
- Reading the interference in terms of superimposition (2) and phase difference (1). Students identify these as fundamental elements but a deeper justification is missing (even a qualitative one).
- Identifying the parameters which determines the pattern, that are: number of slits (3), width of the slits (1), slits-screen distance (1), and wavelength (1). Only in the case of the width of the slit the dependence is specified.

18/32 answers to change the source, but 14/32 students suggest to change grating (5), the width of the slit (4), or the distance grating-screen (5), evidencing difficulties in recognizing the role of these parameters (as emerged in Ivanjek *et al.* 2015). Considering the final reports these difficulties don't appear, and moreover a rich pattern of aspects are mentioned or deeper discussed by students as resumed in the following table (including only main aspects).

Table 1. Mentioned aspects from students in their reports

	DESCRIBING THE LIGHT PATTERN				HUYGENS-FRESNEL			INTERFERENCE			PARAMETERS INVOLVED			
STUDENT	SYMMETRY	ANGULAR DISTRIBUTION	FRINGES	CENTRAL MAX	Just MENTION	SECONDARY SOURCES	WAVEFRONT AS ENVELOP	Just MENTION	SUPERIMPOSITION	PHASE DIFFERENCES	NUMBER OF SLITS	WIDTH OF SLIT	DISTANCE SLIT-SCREEN	WAVELENGTH
1	✔		✔	✔	✔			✔						
2	✔		✔	✔	✔	✔	✔	✔	✔		✔		✔	
3	✔	✔	✔	✔	✔	✔	✔	✔			✔	✔		✔
4	✔	✔			✔	✔		✔	✔	✔	✔			

4. Main results

From the analysis of the final reports, despite the exiguity of the sample, it has been possible to observe that following the proposed path student acquire confidence with the role of the parameters involved in the phenomenology, and activate the construction of a new interpretation of the phenomenology. Every student recognized the need to change from a ray model to a wave one, using the Huygens principle to describe the phenomenology. The description remains however on a qualitative plan, regarding both the principles (Huygens-Fresnel and superposition) and the dependencies from various parameters (number and width of the slits, wavelength…).

5. Conclusions

Starting from a previous didactical proposal about single-slit diffraction, a path concerning diffraction from multiple slits has been developed. It takes into account the learning results emerged from the experimentations concerning the single-slit diffraction educational proposal that aims to obtain phenomenological laws describing the phenomenon starting from the analysis of experimental data.

- Results from pilot studies conducted with upper-secondary school students and freshmen in biotechnology confirm the hypothesis that the path about multiple-slits diffraction helps the acquisition of a new modality in which students see phenomena related to the wave nature of light. Students' personal involvement in phenomenological analyses, as the one described here, offer them the opportunity to experience the way in which experimental research is carried out.
- Students gain ownership in focusing the relevant quantities and laws describing the phenomenon, and the role of the various elements. Such activities, moreover, activate the relationships between the phenomenon and the interpretative models.
- Our future work will be in two directions: to adjust the proposed path on the basis of the results emerged, enriching it with tests, tutorials, didactical materials and activities in modeling, and to experiment the path with students, to deepen its implications for learning, and teachers, for their professional development.

References

Corni F., Mascellani V., Mazzega E., Michelini M. & Ottaviani G. (1993) A simple on-line system employed in diffraction experiments. In *Light and Information* (Pereira L.C., Ferreira J.A. & Lopes H.A., eds.), Braga: Univ. do Minho, pp. 381-388.

Gervasio M. & Michelini M. (2009) Lucegrafo. A USB Data Acquisition System. In *MPTL14* (Lamboune B. *et al.*, eds.). Proc. http://www.fisica.uniud.it/URDF/ mptl14/contents.htm.

Ivanjek L., Shaffer P.S., McDermott L.C., Planinic M. & Veza D. (2015a) Research as a guide for curriculum development: An example from introductory spectroscopy. I. Identifying student difficulties with atomic emission spectra, *American Journal of Physics* **83**(1), 85-90.

Ivanjek L., Shaffer P.S., McDermott L.C., Planinic M. & Veza D. (2015b) Research as a guide for curriculum development: An example from introductory atomic spectroscopy. II. Addressing student difficulties with atomic emission spectra, *American Journal of Physics* **83**(2), 171-178.

Michelini M. & Stefanel A. (2015) Upper secondary students face optical diffraction using simple experiments and on-line measurements. In *FFP14* (Triay R. eds.).

Santi L., Mazzega E. & Michelini M. (1993) Understand Interference by means of computer model. In *Light & Information* (Pereira L.C. *et al.*, eds.), Braga: Univ. do Minho, pp. 372-380.

Vosniadou S. (2008) International handbook of research on conceptual change. New York: Routledge.

IOP Conf. Series: Journal of Physics: Conf. Series **1076** (2018) 012012 doi:10.1088/1742-6596/1076/1/012012

Elementary optics lab with smartphones and instant messaging

Gerhard Rath

University of Graz

Abstract. Smartphones were integrated in an introductory optics course with elements of inquiry based learning. The aim was for the students to gain knowledge in basic optics, to exercise fundamental scientific methods, and to use the student's mobile devices in meaningful ways. Beside the LED light, the camera and some apps, in particular WhatsApp was used as a tool for communication and exchange. A small evaluation on the usage of instant messaging for physics lessons brought good results and lead to some suggestions for the integration of WhatsApp into physics lessons.

1. Introduction

How can we integrate smartphones low-key and easy into physics lessons? This question follows the fact, that we teach students of the so called "Generation Z", who grow up in a digital world, using mobiles and tablets in a natural way within their daily lives (Scholz 2014). On the one hand, the usage of smartphones in secondary schools is under debate. In some schools they are even totally forbidden. On the other hand, a lot of articles and publications, such as the iPhysicsLabs series in *The Physics Teacher* (e.g. Thoms *et al.* 2013), promote working with mobiles in physics teaching and learning, mostly within experiments.

Our approach aimed at using mobile devices in a similar way adolescents do in their everyday lives. It was not announced to the students as a unique or exceptional situation, but we integrated mobile phones as naturally and simple as possible. This particularly concerns the usage of WhatsApp as instant messaging service for communication and data exchange. There are only a few studies about the integration of this tool into teaching and learning activities (Bouhnik & Deaschen 2014, Cifuentes & Lents 2010, Rambe & Bere 2013). Therefore we concentrated our evaluation on this aspect. Within a diploma thesis about instant messaging services in physics teaching a group of 15 students was surveyed using questionnaires.

Regarding content the field of optics seemed proper for this approach. It was adapted after some published contributions to optics (Haagen-Schuetzenhoefer 2017, Colicchia & Wiesner 2015) and also followed the work of H. Muckenfuss. We only needed a small number of apps as the most work was done with the internal camera and the LED light of the mobiles. Most of the pictures in this article were taken by the students with their mobiles. The didactical path started with students' conceptions on "How do we see?". This was followed by investigations of the behavior of light, step by step going to the level of a geometrical model. At last, we came back to explanations of daily experience and technical applications using this model, mainly focusing on the field of image formation. Some aspects of the Nature of Science were consciously integrated, for instance reduction and abstraction.

All of that was embedded in real time communication. Using WhatsApp, pictures and screenshots were exchanged and discussed, including some observations at home. The lessons were held within an interdisciplinary subject called Science Lab, a combination of biology, chemistry and physics lab, and one semester of an interdisciplinary project. The physics part contained an introductory optics course in four groups, each of about 15 students, in 8th grade. The subject is mainly organized as lab work in small groups with student's experiments, using some elements of Inquiry Based Learning, mostly on guided levels (1-2 following Blanchard) (Blanchard *et al.* 2010).

2. Introductory optics

How do we see?

We started with a simple exercise: You see a sheet of paper lying in front of you. How does an image of the sheet come into your brain? The student's drawings in Fig. 1 show a wide range of conceptions. In total, about half of them already used physical models, but some students were still thinking that rays were coming out of the eyes. After a discussion concerning optical illusions we came up with a simple, physical model of seeing.

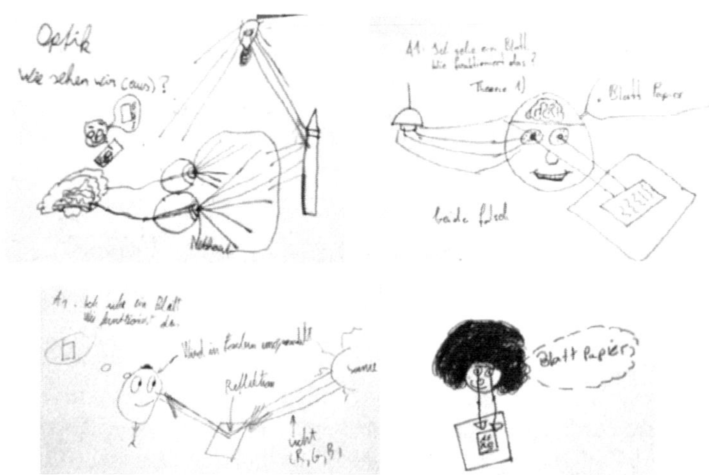

Fig. 1. Drawings of students: How does an image of a sheet of paper come into your brain?

Fingers behind a glass of water

The first experiment discussed a rather daily observation. What do fingers behind a glass of water look like, seen inclined from above? Most of the student's assumptions were wrong. It was very surprising for them that sometimes the fingers could not at all be seen through the glass of water. A lot of questions occurred as the situation is in fact very complex, when trying to explain it with our model. Here we used the smartphone camera for the first time (Fig. 2). The pictures could be exchanged quickly and provided suggestions for new observations. In addition, they showed that the phenomenon is not an optical illusion created by our brain, but must be a physical phenomenon, when a technical tool can fix it.

Fig. 2. Fingers behind a glass of water

From seeing to light

So we had to change the focus from our perception to the behavior of light passing through the glass. Using the phone's light we got fascinating pictures (Fig. 3). Unfortunately the situation stayed complicated, as a lot of possible arrangements could be made. The results depended on many variables, such as the angle of incidence, the glass or the water level.

Fig. 3. The smartphones LED passes a glass of water

Forcing the light

In order to create measurable situations, which are needed for scientific evidence, we bundled the light into thinner rays (Fig. 4). The left picture realized this with the light of the phone combined with a lens and a slit. In the right picture we can see the next step of reduction: laser beams and geometrical objects of glass. This setup enabled the students to verify the basic laws of reflection and refraction, which made us reach an abstract level of geometrical and mathematical models.

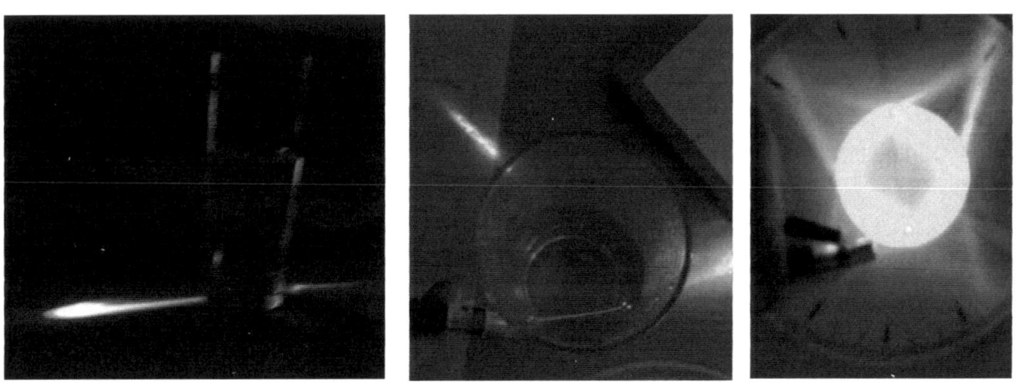

Fig. 4. Left: A "beam" generated by the phones LED with lens and slit passes a glass of water. Middle: Laserbeam in water; Right: Laserbeam in semispheres of glass

Image formation

Staying on this abstract level, we investigated the image formation of lenses. Phone displays were used as light emitting objects, as the student's picture shows (Fig. 5). The relation between size and distance of the object and the image as well as the lens equation could be estimated. On this level we also used the smartphone as a simulation tool to support the image construction, using the app "Ray optics"[8].

[8] Ray Optics for Android: https://play.google.com/store/apps/details?id=com.shakti.rayoptics

Fig. 5. Image formation: A smartphone display (left) is used as light emitting object, the convex lens in the center generates a picture on the screen to the right

Water droplet lens

A nice experiment for further inquiry followed: Place your phone on the table with the front camera on. Approach an object to determine how close you can put it to the camera while still getting a sharp image and what maximum size you can get (Fig. 6). As a next step, put a water droplet on the camera lens. What will happen? With that experiment we returned to the level of explaining the physical function of the human eye, or the camera as a technical application, using optical phenomena.

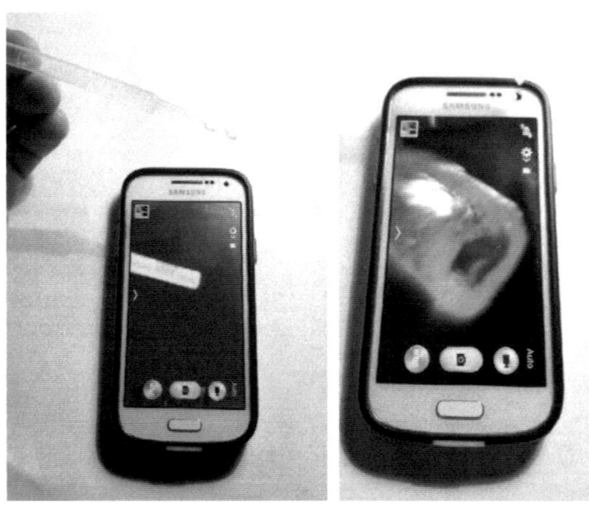

Fig. 6. Left: Approaching a pipette to the front camera. Right: With a droplet of water on the front lens we get an enlarged picture

Colors

The phone's display can create colors through a superposition of red, green and blue light. Surprisingly we can even measure colors, not in wavelengths, but in coordinates of color spaces (Bengtsson *et al.* 2014). The most suitable color space for that is the HSV space, with hue (H) as an angle from zero to 360 degree, representing the light spectrum. Measuring colors is a challenge and using the recommendable app "Color Grab[9]" one can learn that colors not only depend on the object, but also on the surrounding light (Fig. 7).

[9] Color Grab for Android: https://play.google.com/store/apps/details?id=com.loomatix.colorgrab.

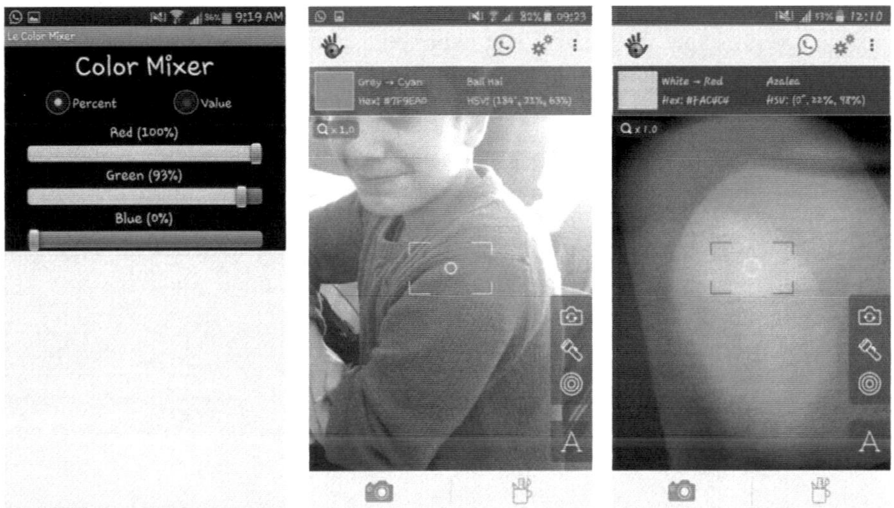

Fig. 7. Left: The app "RGB Color Mixer"[10] generates yellow as a superposition of red and blue. Middle: Color of a piece of clothing (Color Grab) Right: Mixing light (red, green, blue) of lamps with estimation of the colors (Color Grab)

For all units WhatsApp was mainly used as a tool to send the pictures to the group instantly and to discuss them, in addition to ordinary lab minutes.

3. Evaluation of the use of WhatsApp

There had been no systematical evaluation about the effectiveness of the experiments. The outcome (pictures, discussions) showed a lot of successful activities, the feedback of the students was generally positive. Both had been communicated using WhatsApp, so we decided to perform a small evaluation concerning the usage of this tool in one of the classes (14 pupils). It was part of a study related to the impact of WhatsApp on doing homework, a diploma thesis at the University of Graz (Sorschag, 2016). The answers to the open question about the perception of WhatsApp were categorized. They showed acceptance for using WhatsApp for science classes (Fig. 8).

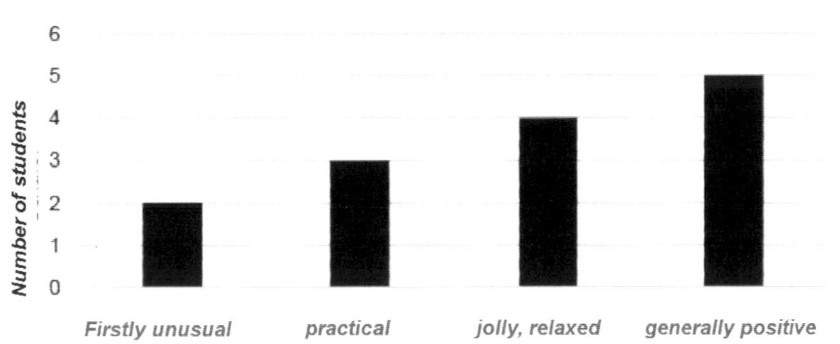

Fig. 8. Answers to the question "*How did you perceive the communication with your teacher via WhatsApp?*"

[10] RGB Color Mixer for Android: https://play.google.com/store/apps/details?id=com.FlyingSolo.src

Fig. 9 shows a very high acceptance of the work with WhatsApp in the questioned group. Other groups had different results, but in the long run experience showed that the majority appreciates the extended communication. Nevertheless, there were always some students who did not like it or were not able to use WhatsApp at all.

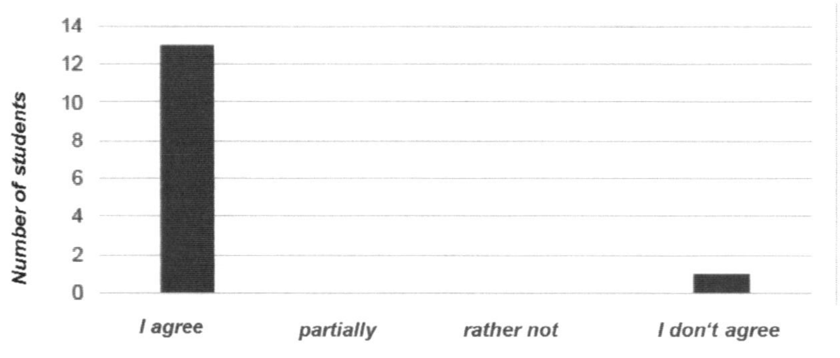

Fig. 9. Answers to the argument "*I would like to use WhatsApp again in physics class*"

4. Conclusion

It showed that student's smartphones can be a useful tool to support the teaching of elementary optics, when integrated in a natural way and mainly used in practical work. Further research could ask for the impact on the learning outcome or other possible consequences of the smartphone usage. Concerning WhatsApp we can point out some positive aspects:

- The real time communication can immensely support lab work.
- The students are absolutely familiar with this tool.
- We may open a new access point to physics, using this familiarity.
- WhatsApp is suitable for multimedia – pictures, text, videos and sound.
- It economically uses the volume of transferred data, which is important when sending pictures and videos.

On the other hand we could locate some problems:

- The usage causes a mixture between private life and work, mainly because you need to make your phone number more or less public in order to join groups.
- Skype and Facebook offer versions for the computer without the need of a mobile phone. WhatsApp has a PC version too, but this only works when connected to the phone.
- The linear chronology is sometimes impractical. Messages follow in real time and you do not have any structure or grouping.

Acknowledgements

I want to thankfully mention the students of the BRG Kepler in Graz doing the practical work, in particular the photographs in this article. Many thanks to Julia Sorschag, who did the questionnaires within her diploma thesis, and Angelika Renz for the language corrections.

References

Bengtsson D., Jonas L., Los M., Montangero M. & Szabo M. (2014) How Deep is Your Blue? Coloured Chemistry with Smartphones. *iStage2: Smartphones in Science teaching*. Science on Stage Germany, Berlin.

Blanchard M.R., Southerland S., Osborne J., Sampson V., Annetta L. & Granger E. (2010) Is inquiry possible in light of accountability? *Science Education* **94**(4), 577-616.

Bouhnik D. & Deshen M. (2014) WhatsApp goes to school: Mobile instant messaging between teachers and students. *Journal of Information Technology Education: Research* **13**, 217-231.

Cifuentes O.E. & Lents N.H. (2010) Increasing Student-Teacher Interactions at an Urban Commuter Campus through Instant Messaging and Online Office Hours. *Electronic Journal of Science Education* **14**(1), 1-13.

Colicchia G. & Wiesner H. (2015) Looking into the Eye with a Smartphone. *The Physics Teacher* **53**(2), 106-108.

Haagen-Schützenhöfer C. (2017) Development of Research Based Teaching Materials: The Learning Output of a Course for Geometrical Optics for Lower Secondary Students. *Key Competences in Physics Teaching and Learning* (Dębowska E. & Greczyło T., eds.), GIREP EPEC Conference 2015, Wroclaw, Springer.

Rambe R. & Bere A. (2013) Using mobile instant messaging to leverage learner participation and transform pedagogy at a South African University of Technology. *British Journal of Educational Technology* **44**(4), 544-561.

Scholz C. (2014) *Generation Z.* Weinheim: Wiley-VCH.

Sorschag J. (2016) *WhatsApp, Prof? Instant Messaging Dienste im Physikunterricht.* University of Graz.

Thoms L., Colicchia G. & Girwidz R. (2013) Color Reproduction with a Smartphone. *The Physics Teach*er **51**, 440-441.

IOP Conf. Series: Journal of Physics: Conf. Series **1076** (2018) 012013 doi:10.1088/1742-6596/1076/1/012013

Study of oscillatory motion using smartphones and tracker software

Antonio Amoroso[1], Marta Rinaudo[2]

[1]University of Torino and INFN Torino
[2]University of Torino

Abstract. The purpose of this contribution is to propose a workshop for acquiring data through the use of a smartphone, not necessarily high end, equipped with a camera and the required application installed. With a smartphone and the free Physics Toolbox application, it is possible to study the oscillatory motion of both simple and vertical pendulums. This is possible by measuring the components of acceleration on the three axes by means of a reading from the internal accelerometers of the smartphone. The vertical motion of the pendulum brings to light the concept of elasticity. Thanks to the Tracker software it is possible to study the damped oscillatory motion of a moving body on a semi-circular drive through the video analysis of the motion itself. Tracker is a free video analysis and modelling tool built on the Open Source Physics Java framework. Given the simplicity of the acquisition procedure and data analysis it is possible to study the motion of circular bodies of different sizes and materials in a short amount of time. The activity relating the vertical motion of the pendulum study also allows for the comparison between the classical procedure, using a stopwatch and both proposals procedures, using smartphones and Tracker. The analysis of subsequent data acquisition must be carried out using standard spreadsheet calculations by introducing the concept of error and the techniques to make it as small as possible. The use of the new widespread technologies allows students to approach the physical concepts even outside of a classic laboratory. The ease of use and low cost allow any school environment to propose activities to the entire class by offering more contemporary workstations and organizing groups of a few students.

1. Introduction

The course that has been proposed is a part of the course for the teaching qualification, designed as a series of laboratory activities suited to secondary school. Often in schools there isn't an equipped place to use as a laboratory. It is therefore necessary to think some alternative experimental activities carried out with everyday equipment in order to offer a good example that teacher can replicate in the classroom. The use of the software "Tracker", smartphones and tablets are a possible solution of these problems. Tracker is a free video analysis and modelling tool built on the Open Source Physics Java framework (only for PC).

By measuring the acceleration components on the three axes by the internal accelerometer of the telephone, it is possible to study the oscillatory motion of a simple pendulum and a vertical pendulum. The study of motion of vertical pendulum allows to introduce the concept of elasticity. It is also possible to study the damped oscillatory motion of a moving body on a semicircular guide.

Experience of simple pendulum

Using a smartphone as a mass it is possible to build a pendulum (Vogt & Kuhn 2012) of length l and measuring the period of oscillation by acquiring the data through the application "Physics Toolbox accelerometer" (Fig. 1). This App is free downloadable both for Android and iOS.

The software acquires the values of the components of acceleration on three axes, records them in a format compatible with the commercial programs for data processing and displays a screen trend graph. Analyzing the data is possible to study the relationships between the oscillation period and pseudo period depending on the length of the pendulum (Fig. 2).

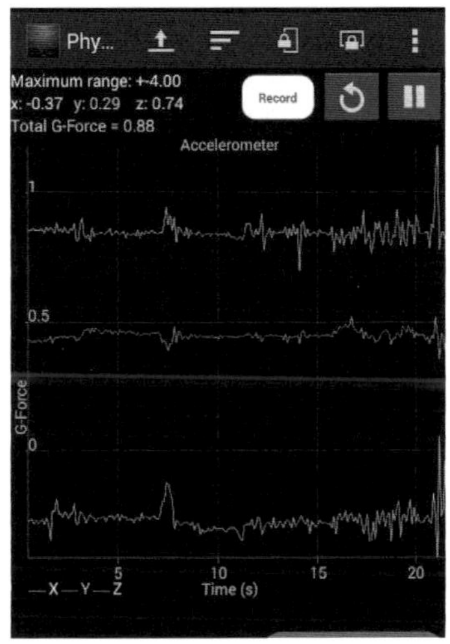

Fig. 1. Screenshot of "Physics Toolbox"

Observing and evaluating the damping factor due to friction may be estimated how many oscillations the period remains constant.

By changing the angle of oscillation, the length of the wire and the hanging mass it is possible to study the dependence of the oscillation period by one of the three variables.

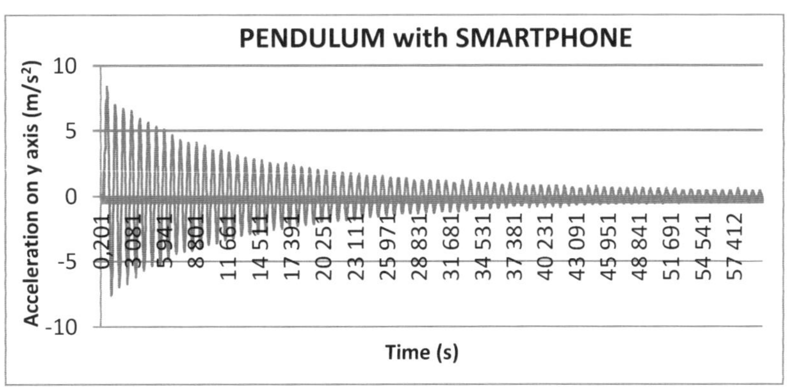

Fig. 2. Data of the pendulum acquired with Physics Toolbox

The results obtained with Physics Toolbox can be compared with the data extracted from the analysis of the motion of the pendulum.

The Tracker software allows the student to capture video and select the position of the mass of the pendulum with respect to the equilibrium point at fixed time intervals (Fig. 3). First of all, it is necessary to make a video of the physics phenomena with the smartphone camera and to choose a calibration length in the video that the program will serve as a reference unit. Then set and the frames interval in which the program will have to identify location of the object and select a point on the object. The program automatically goes to the next selected frame in which to repeat the reference point selection procedure (always the same point).

IOP Conf. Series: Journal of Physics: Conf. Series **1076** (2018) 012013 doi:10.1088/1742-6596/1076/1/012013

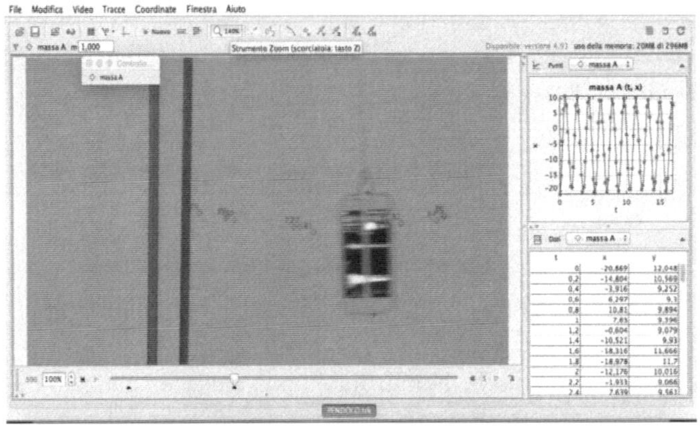

Fig. 3. Screenshot of Tracker for the pendulum

The Tracker software allows the view of motion of the pendulum and, in parallel, the construction of the point-by-point graph. In particular, in our case it is interesting to separately view the acceleration performance and its three components. The acquired data can be exported in a format compatible with leading software for a statistical analysis

Experience of vertical pendulum

Using a smartphone as a mass and a spring with a constant of elasticity k, it is possible to build a vertical pendulum (Kuhn & Vogt 2012). The law that describes the motion is harmonic damped along the vertical axis:

$$x(t) = A\cos(\omega t + \Phi)$$

through the measurement of the oscillation period it is possible to obtain the value of constant of elasticity k of the spring.

$$T = 2\pi\sqrt{\frac{m}{k}}$$

The experimental measurement is carried out, similar to the pendulum one, using your smartphone as a mass m and "Physics Toolbox accelerometer" to acquire the period of oscillation of the vertical pendulum: the use of a smartphone as mass is not the best choice to study the motion of vertical pendulum (Fig. 4 – it's very difficult to get perfect vertical oscillations due to the size and mass of the smartphone).

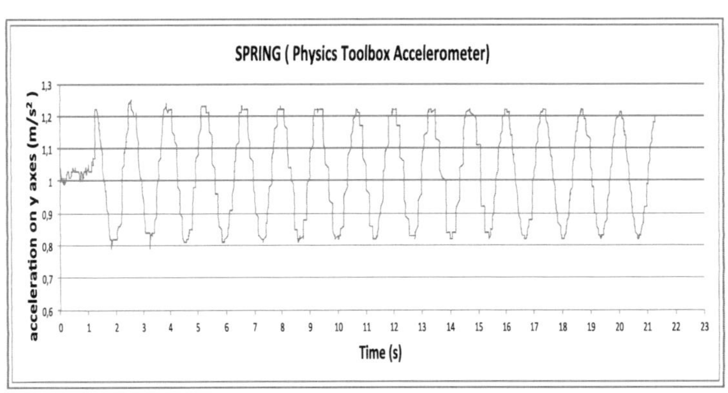

Fig. 4. Data of the vertical pendulum acquired with Physics Toolbox

In this case the use of the Tracker program is more effective (see Fig. 6). Through the video analysis of the oscillations, the experimental period can be obtained and then the experimental value of k.

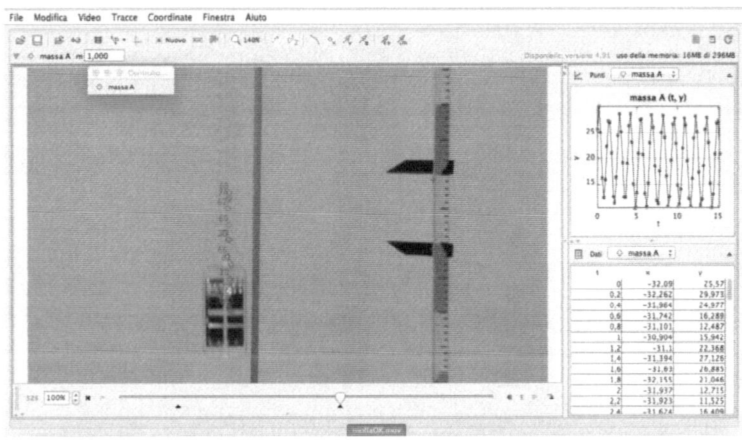

Fig. 5. Screenshot of Tracker for the vertical pendulum

Damped oscillatory motion

Using a rigid semi-circular guide it is possible to view the damped oscillatory motion of an object free to move on the guide itself. The rear fender of a bicycle and a styrofoam ball (Fig. 6) are sufficient to obtain an adequate experimental setup to the proposed study; and the size of the ball was chosen so as to be always easily visible in any frame of the video analyzed with the Tracker program, the light-weight material (polystyrene) instead allows an adequate number of oscillations. Using objects of different sizes and materials, it is possible to study (for short time intervals) their oscillatory motion and the relation between amplitude and speed.

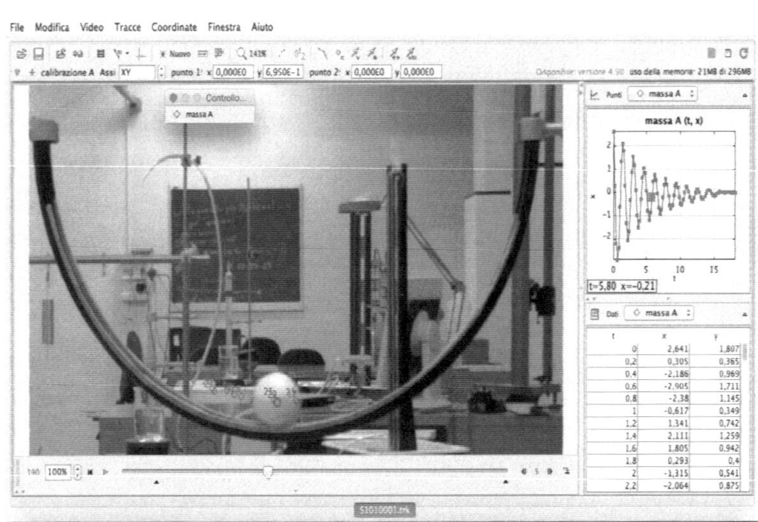

Fig. 6. Screenshot of Tracker for the damped oscillation

This experience allows to introduce, experimentally, the concept of damping e, of gravitational potential energy, and to study the dependence of the oscillation amplitude oscillation period itself (Fig. 7).

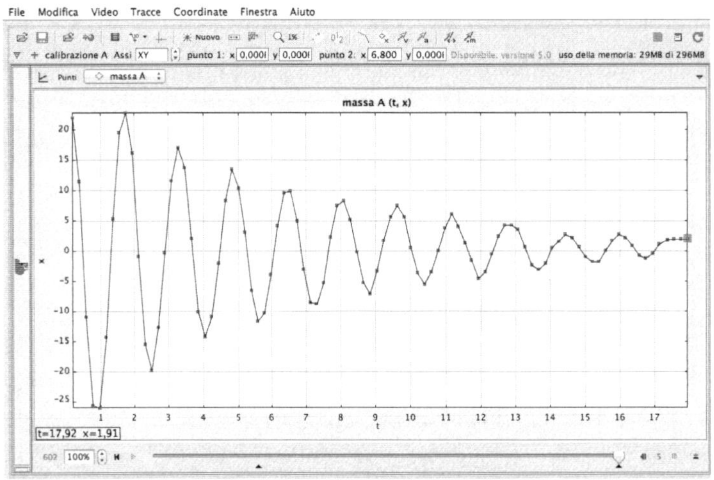

Fig. 7. Graphic obtained with Tracker

2. Conclusions

Proposals experiences are an example of how the use of new technologies widely deployed to enable students to approach the physical concepts, even outside of a classical laboratory. The ease of use and low cost allow any school environment to propose activities to the whole class with the possibility to set up multiple workstations and groups made by few students. The data analysis can be carried out using a common spreadsheet associating to the physical quantities the related errors.

The proposed software can be used to prepare a wide range of experiences allowing the teacher to show practical applications of physical theories discussed in class.

References

Vogt P., Kuhn J. (2012) Analyzing simple pendulum phenomena with a smartphone acceleration sensor, *The Physics Teach*er **50**, 439-440.

Kuhn J., Vogt P. (2012) Analyzing spring pendulum phenomena with a smart-phone acceleration sensor, *The Physics Teach*er **50**, 504-505.

Enjoy physics classes with your own devices

Beata Jarosievitz

Dennis Gabor College

Abstract. Introducing the latest educational technology trends has accelerated enormously in the past few years; therefore, the use of personal devices, especially smart phones, tablets, laptops has increased considerably also in the educational processes. Many researchers used M-learning for different purposes (Hsu & Ching, 2013), but only a few of them used it for physics teaching experiments (Crompton, Burke, Gregory & Gräbe, 2016; Jarosievitz, 2016; Kuhn & Vogt, 2013). The use of M-learning devices in experiments is based on the rich set of built-in sensors in smart phones (Kuhn & Vogt 2013; Staacks, 2016). However, beside the devices themselves, also free applications and teachers' (instructors') expertise is required.
In this work some meaningful use of M-learning during physics lectures will be presented. Some of the key terms describing the quality of the measurement have also been discussed with the students, and will be presented here. After the conclusion of the measurements, students used their own devices as clickers, and answered the questions synchronously and anonymously through an on-line assessment system.

1. Introduction

Based on the conclusions of my previous research activity made for my PhD study (Jarosievitz, 2005), and analysing different papers afterwards, it has become clear that the Physics education is in crisis in many parts of the world (OECD 2000, 2001, 2005; Sadowska & Kamińska 2010; Guido 2013; Grazier 2016).

It is a very interesting question why students do not like Physics, and how we can turn back their attitude in a positive way.

This crisis in Physics education is usually deepest in the primary and secondary sector. One of the causes of this crisis might be that in primary and secondary schools the number of Physics classes has been extremely decreased in several countries – also in Hungary. In parallel to this, the number of physics teachers has also decreased, so that those who are still working should teach 21-26 classes a week.

During the reduced number of hours the majority of the students cannot do any hands-on experiment, students became passive learners, they do not do experimental work during the classes, and therefore without experiments it is much harder to understand well the physics phenomena or topic.

Many good schools in Hungary are mainly concentrating only to problem solving during the Physics classes, do not let students do any experiments. The main reason is that the teachers teaching more than 21 classes per week, do not have enough time for setting up the experiments – even if they would have well-equipped laboratories, and they do not have technicians, who can help them with the preparation of the experiments.

In the 21st century the pure traditional physics classes are not good enough for attracting the students' focus to the lectures. Many of the students may not enjoy the series of lessons (Turner 2015).

However, students have new tools in their hand, which can be used, and touching, doing something much more interactive attracts them better, instead of listening to the teacher's explication.

The higher education level is not exempt of that crisis either, because of the decreased quality of the students leaving the secondary schools and entering in the university or college.

Some universities or colleges do not have students' laboratories at all (at least here in Hungary), or the students' laboratory (experimental) work is not included in their curricula (e.g: BSc in Engineering Information Technology).

Students studying in these places do not carry out any experimental work, and they are not familiar either with the measurements, or the key terms which describe the measurements' quality. Students in these places are not involved at all in experiments (Jarosievitz 2011).

Qualitatively analysing some informal feedback of my students, I have found that the physics lectures should be made more colourful, attractive, interesting and interactive, especially in the first year of the university or college. Many researchers and teachers advised different methods and activities to increase the recognition of the Physics education (Bae & Kim 2013, Graham 2017).

Among these new trends the use of personal devices (BYOD), or other parallel interactive activities can be used, like Researcher's Night (Jarosievitz 2012), Science on Stage Festivals *http://www.science-on-stage.eu etc*. Students using their own devices can take part interactively in the courses, and we do hope that accessing and sharing information with these mobile devices will motivate them, and will change their attitude towards science in a positive way.

If we let our students to leave the university or college without any skills in designing and performing hands-on experiments, they will become only theoretical professionals, and they will never be able to do any practical work, or inquiry based investigations.

We should not let our students to become digitally illiterate; therefore we must act immediately introducing using of different constructivist based methods (Grazier 2016).

2. M-learning devices used during the lecture

The definition of M-learning by Wexler: "any activity that allows individuals to be more productive when consuming, interacting with or creating information mediated through a compact portable digital device that the individual carries on a regular basis, has reliable connectivity and fits in a pocket or purse" (Wexler *et al.* 2008).

Seeing the difficulties mentioned in the introduction (e.g. lack of experimental work from curricula), I decided to introduce the use of innovation based experience of M-learning. I am aware that M-learning cannot replace authentic physics experiments, but they still can help students to predict and to explain physics phenomena, to calculate the accuracy, error, etc. They are also excellent tools for using cooperative learning, to work together in small groups on a structured activity, and for thinking first before acting.

This work has been done as part of a research program designed for first-year students in Physics at the Dennis Gabor College from Budapest, Hungary. This teaching experiment with mobile devices involved 188 students who were enrolled in my course.

At the outset of the course 43% of all 188 students (20.7% female; 79.3% male of them) filled in the general questionnaire (31 questions of different types) that was sent to them via the internal communication system. Only a few of selected questions and answers are presented in this paper, why the goal of this paper is not related to analysing the student's attitude to their study. I made this study only to have some idea about the group involved in the experiment.

Analysing the responses of the students made before we started our lectures, I found their reactions that 24.4% are not really interested in using their own devices for physics measurements. 75.6% students answer, that they positively motivated to see some new ideas, implementation, and they looked forward to taking the measurements with their own devices.

I was also curious about what purposes the students used the new high-tech devices for, if they had one. Examining the responses I found that my students used their devices mostly for communication, and my results are very similar to other results made in a completely different part of the world (Foti 2014).

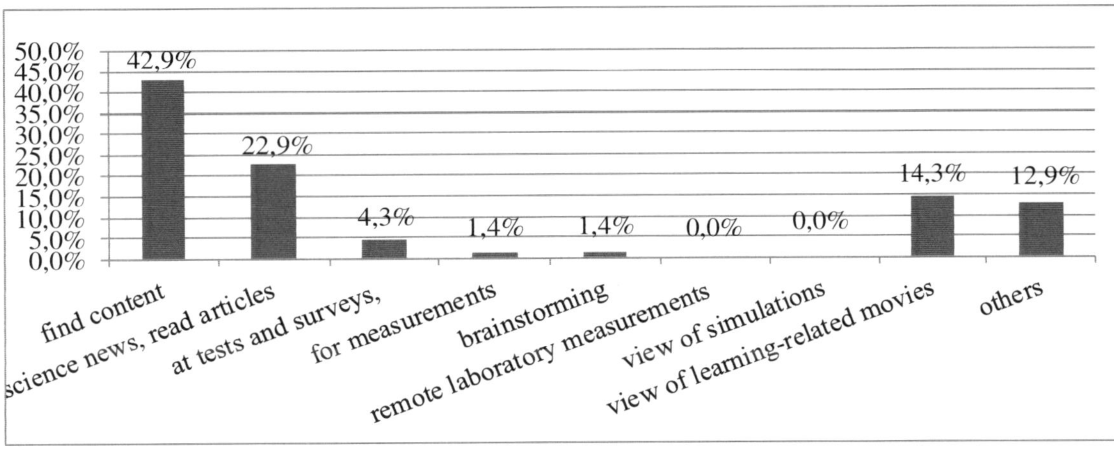

Fig. 1. Portable devices used by students, for different purposes

3. Working hypothesis

"A working hypothesis is a hypothesis that is provisionally accepted as a basis for further research in the hope that a tenable theory will be produced, even if the hypothesis ultimately fails" (Wikipedia 2017).

Before starting my innovative pilot experience using portable devices (smart phones, tablets, laptops) for real physics experiments in Physics teaching classes, I have formulated the following working hypotheses:

Students turning to their neighbours and using their own devices will:

- find the physics lectures more enjoyable;
- be able to take different measurements;
- be able to discuss the measured values;
- increase their interest regarding to physics and the sciences,
- increase students' problem solving and computational competencies, and also help to understand better the corresponding physics phenomena.

4. Real measurements done with personal devices

4.1. Aim of the Measurements

Here I present only two real measurements performed with portable (Android) mobile devices. The aim of both measurements was a classical one: to determine the value of the acceleration due to gravity without using the gravity sensor accelerometer of the device. Acceleration is quantified in the SI unit system by m/s^2.

4.2. Methods used for Activity

The experiment was performed with 36 students, who took part in the 3-hour lectures.

Before starting the activity first I grouped the students into informal cooperative, heterogeneous, learning groups (Brame & Biel 2015). Informal group-work consisted of the following:

4.2.1. Think-pair-share (T-P-S)

"T-P-S technique is designed to encourage students to share and discuss ideas around a particular topic, issue or problem" (Lyman 1981).

As "instructor of the group" I have given some instructions to my students and let them discuss and answer my questions formulated specially for them.

Questions

- If we measure the "g" value using different methods, what do you think, do we get exactly the same results as published already?
- Will the value of "g" be the same everywhere in the world?

First each student considered the questions, then they discussed in pairs, and finally changed their place and started to discuss their responses with other students from different groups. Every group shared their responses with all other participants.

The following Fig. 2 visualises the arrangement of the students in different working groups.

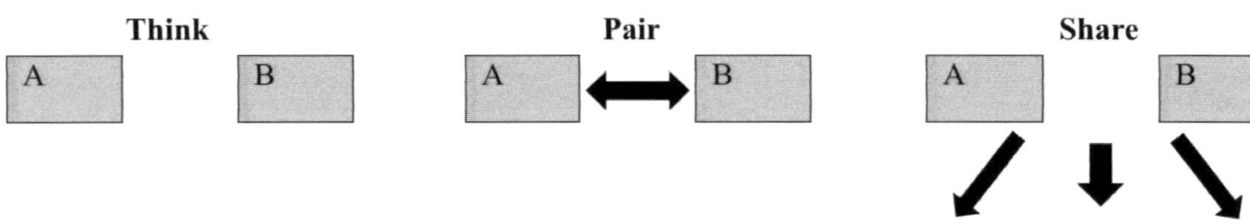

Fig. 2. Examples of informal cooperative learning activities Think-pairs-share (adapted from Brame & Biel 2015)

4.2.2. Peer instruction

Peer instruction is an evidence-based, interactive teaching method successfully pioneered and popularized by Harvard Professor Eric Mazur in the early 1990s (Mazur 1997).

The new technique known as peer-instruction can be used with and without technology (e.g., clickers). Peer Instruction encourages students to critically think through the arguments being developed, and to discuss their ideas with their neighbours.

This activity makes students better involved in their own learning, and focuses their attention on underlying concepts.

Since we did not have any clickers (special devices), my students used their own devices (tablet, smart phones, laptops).

Students answer different type of questions online, that means the students should use the network of the college, or university. This way the expensive "clickers" were replaced by M-learning devices, especially smart phones.

The main aim of the use of the Classroom Response Systems "Clickers" (CRS) was to promote active student engagement during a lecture, and promote discussion and collaboration among students.

All of my students got a QR code (with the direct link to the online poll: *http://www.socrative.com* and room number: **f862bde5** used) already prepared in advance. Students first need to scan their QR code. If their device could not do it by default, they had to first download and setup an appropriate app into their own devices. As a first step students scanned the QR code and logged in as a students to the free online poll:

Fig. 3. QR code prepared for students with direct link to the online poll

Students without devices had to turn to their neighbours and discussed all questions and answers, before they voted in pair. Every answer was anonym and shared with all participants using the projector. This approach was well-received by all students.

Using the students' own M-learning devices I got very good real-time information regarding their knowledge, while students shared their understanding by answering anonymously the formative assessment questions in a variety of formats: quizzes, quick question polls, exit tickets and space.

4.2.3. Jigsaw

Jigsaw is a method of organizing classroom activity that makes students dependent on each other to succeed. It breaks classes into groups.

Students work in a team of four to become "experts". In our case each group of four students worked on the followings:

- setting up the experiment,
- take the measurements,
- work on data analysis,
- do the graphical representation of the measured values,
- prepare the final report,
- work on the conclusions,
- checking the validity of the hypothesis defined in the beginning.

Each student of each group became an "expert". Then they changed their places and the whole class was rearranged, forming new groups, keeping one "expert" member from previous groups. The new group was informed and taught by the "expert" member. See Figs 4, 5.

A	A
A	A

B	B
B	B

C	C
C	C

D	D
D	D

Fig. 4. Students became expert on the task that they were doing

A	B
C	D

A	D
B	C

D	C
A	B

C	B
D	A

Fig. 5. Rearranged class to allow peer-to peer instructions

4.3. The Real Measurements Done

4.3.1. Measurement 1

This measurement has been carried out with every small group, where one of the "experts" gave the instructions based on the previously described method.

One of the methods to determine "g" based on the observation of a falling ball in two dimensions (parabolic throw) is presented below. This measurement has been done with M-learning devices, with the new method (Juhász *et al.* 2015), instead of the usually employed free-fall (one-dimensional) methods.

For the first measurement we need the following materials: steel ball; ruler; laptop (equipped with a sound card) + microphone + projector; Audacity free program (software programs for sound recording) installed in the laptop (which can be downloaded from: *http://www.audacityteam.org/* site.

(If we do not want to use a laptop and a microphone connected to the laptop, we can directly use our smart phone or tablet with a free app from Google play, called Audio Mix Studio. Using this app exactly the same measurement can be carried out, as can be done using a microphone and laptop. Actually students used their laptops for the measurements, while we shared our screens online with the audience. The laptop screen was projected to the observers.

After setting up the experiment, someone had to hit the steel ball which starts to move on the clean table. Group of the students, as observers followed the little steel ball motion, which was moving on the clean surface (table) and after a time it fell down, and hit the surface of the floor.

During the falling time we did not hear any sound. The sound file was recorded during the whole motion of the steel ball.

Students repeated the measurement 5 times, and then started to analyse the recorded sound file.

Examining the sound file each group was extremely curious to determine the exact time when we did not hear any sound. The period of time when we cannot hear any noise is actually the free fall time.

Finally, the recorded sound file was analysed and we also had to take into account the noise filtering of the recorded sound file. The recorded sound file of this experiment is plotted in the next figure (Fig. 5).

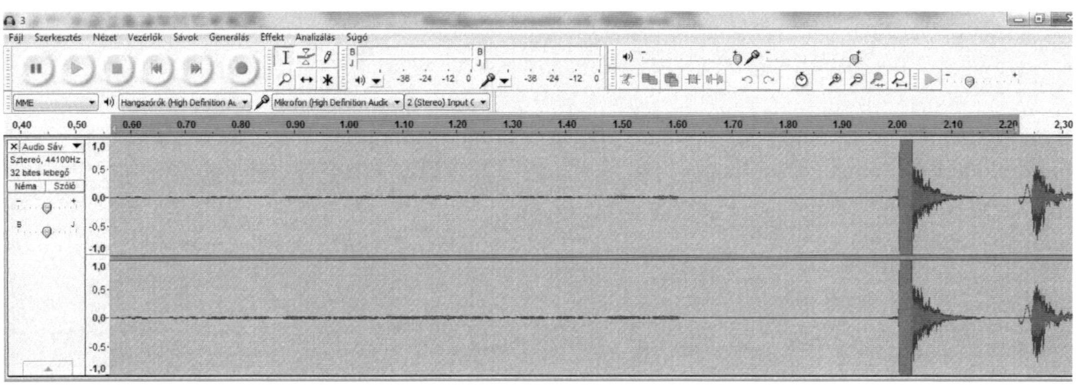

Fig. 6. Recorded sound file after the noise filtering

From the definition we already know the free fall law $h=\frac{1}{2}\cdot g\cdot t^2$
where t is the time of falling, h – height, g – acceleration due to gravity

From the law we can extract the g value, based by the time measured as: $\Rightarrow g=\frac{2\cdot h}{t^2}$

For getting not only the value of the physical quantity but also the standard deviation, this measurement has been done 5 times. Analysing our result we got the following values (table 1).

Table 1. Results after the measurements

No.	**h (m)**	**t (s)**	**t^2 (s)**	**g (m/s²)**	**x_m (mean)**	**(Δx)**	**δx**
1	0.730	0.383	0.147	9.953			
2	0.730	0.386	0.149	9.799			
3	0.730	0.398	0.158	9.217	9.67	0.29	3.0%
4	0.730	0.393	0.154	9.453			
5	0.730	0.383	0.147	9.953			
Average	0.730			9.675			

$$x_m = \frac{\sum x_n}{n} \qquad (\Delta x)^2 = \frac{1}{n-1}\sum_{i=1}^{n}(x - x_m)^2 \qquad \delta x = \left(\frac{\Delta x}{x_m}\right)\cdot 100\%$$

After a common – guided – discussion about the results the students conclude that our measurement gives the following estimation for the gravitational acceleration: $x_m - \Delta x < g < x_m + \Delta x$, which in this case gives $9.35 < g < 10.00$. They compare this result with the "known" value of the acceleration due to the gravity for our latitude: 9.81 m/s^2, and recognize, that this value is well inside the interval. This means to them that this measurement had probably no systematic error – the resulting value is "valid".

"Validity: A measurement is 'valid' if it measures what it is supposed to be measuring. What is measured must also be relevant to the question being investigated" (*http://practicalphysics.org/language-measurements.html*).

Even this measurement has been done with their own devices, students confirm that they enjoyed taking these measurements very much, while they used their everyday devices, and understood the physics "with clicking and doing".

They were very enthusiastic while being involved in this short part of the lecture, and carried out the task actively in each working groups.

4.3.2. Measurement 2

The second measurement had the same aim: to determine the value of the acceleration due to gravity ($g = 9.8$ m/s^2) using another method. The aim of using another method was to show the students another possibility, other free program available, which can be used for physics measurements in case if they do not have a well-equipped laboratory behind.

For this measurement also the informal cooperative learning method was used. Some students set up the experiment, some of them recorded the video file of the action, some of them did the calculations, or wrote the report, and reported their results after some discussion with their neighbours.

Necessary materials used for the experiment were: handball (any large, well visible ball); ruler; smart phone or tablet; laptop; a free video analysis and modelling tool built on the Open Source program called Tracker (can be downloaded from: http://physlets.org/tracker site).

A very brief description of the measurement can also be found on: http://moodle.scientix.eu/course/view.php?id=179 site created by Carlos Cunha, which impressed me a lot. I decided to adapt it to my course in this pilot experiment with some minor changes and additional goals.

Students started the activity and made the "measurement": $y = a \cdot x^2 + b \cdot x + c$ they recorded the motion of the falling ball 5 times. After recording the movements, the harder work started which involved transferring the data to the laptop, and starting the video analysis of each file. The video analysis took longer time, because the program was not a familiar tool used by the students before. In the beginning of the evaluation they had to learn step by step what they should do. The distance-time function of the falling ball is a quadratic function, and the value of "g" should be extracted from this function. The general form of a quadratic function:

For the free fall we have parabolic functions: $y = -\frac{g}{2} \cdot t^2 + v_0 \cdot t + y_0$

$$a = -\frac{g}{2} \qquad \Rightarrow g = -2 \cdot a$$

Table2. Result after the measurements

Files	a	g (m/s^2)	x_m (mean)	(Δx) (st. dev.)	δx
20160311_123159.mp4	4.63	9.26	9.33	0.07	1%
20160311_123217.mp4	4.62	9.24			
20160311_123235.mp4	4.72	9.44			
20160311_123253.mp4	4.67	9.34			
20160311_123316.mp4	4.69	9.38			

where $x_m = \frac{\sum x_n}{n}$ $\quad (\Delta x)^2 = \frac{1}{n-1}\sum_{i=1}^{n}(x - x_m)^2 \quad$ $\delta x = \left(\frac{\Delta x}{x_m}\right)\cdot 100$

After discussing the result the students conclude that our measurement gives the following estimation for the gravitational acceleration: $x_m - \Delta x < g < x_m + \Delta x$, which in this case gives $9.25 < g < 9.42$. Here they also compare the results with the "known" acceleration due to gravity value, and realize that. In this case the value lies well outside the limits defined by this measurement (9.33 ± 0.083 m/s^2). The relative deviation from the generally accepted value is about 5%, whereas our statistical fluctuation is only about 1%. This gives us the opportunity to discuss the possible causes with the students. Here we have the opportunity to mention that according to the probability theory the one-sigma interval contains only about 67% of the measured values. Therefore there is still a substantial probability (about 33%) that the "real" value is outside the one-sigma interval. However, the generally accepted gravitational acceleration is even outside the $x_m \pm 3\cdot\Delta x$ interval (three-sigma interval), and the statistical probability of that should be less than 1%. Therefore we can conclude that this measurement definitely had also some (~ 5%) systematic error: either the "picking" of the exact positions of the ball is not precise enough, since the ball is an extended object, and – additionally – the image of the ball gets more and more blurred as the ball moves faster and faster, making the position-picking more and more difficult and less and less precise, or there is a small timely deviation in the frame rate of the video.

The discussion of these discrepancies was also very informative and useful for the students, since they got a more practical feeling what was the difference between a systematic error and a statistical error.

5. Conclusion

Making anonymous personal interviews with my students, I could conclude that students who participated in these lectures enjoyed very much the use of informal cooperative learning method. They really liked to cooperate with their neighbours "Turn To Your Neighbours" (Mazur 1997).

They felt it was very useful, and it motivated them more to carry out their tasks, to do calculations and to answer the questions.

They also very much liked the idea of bringing and using their M-learning devices to the lecture. Finally they felt that they were better prepared for physics exam, while they had been involved in the measurements. They also understood better the corresponding physics phenomena and laws, while they became active learners during the lectures, and not only passive listeners. M-learning is very useful for learning, for reading and finding relevant content on the Internet, for assessing acquired knowledge and for performing real measurements, therefore it should be implemented much more in higher education activities not only in a pilot experiment.

By making more measurements with M-learning devices one can study much better the students' attitude to physics, and can influence them to increase their motivation and solve problems interactively in cooperation with each other's.

For me it became clear that if we use only the same old strategies, methods and techniques we will never challenge our students', the Z generation's attitude. By using different methods, this pilot activity looks promising to shifts the role of the teacher, and to increase the audience activity.

„The future cannot be predicted, but the future should be invented" (Dennis Gabor).

References

Brame C.J. & Biel R. (2015) Using cooperative learning groups effectively https://cft.vanderbilt.edu/guides-sub-pages/setting-up-and-facilitating-group-work-using-cooperative-learning-groups-effectively/.

Graham E. (2017) Using Smartphones in the Classroom, Tired of telling students to put away their phones? A veteran teacher shares tips for using mobile devices as learning tools. http://www.nea.org/tools/56274.htm.

Crompton H., Burke D., Gregory K.H. & Gräbe C. (2016) The Use of Mobile Learning in Science: A Systematic Review. *Journal of Science Education and Technology* **25**(2), 149-160.

Jarosievitz B. (2005) Opportunities of Information Technology and Multimedia in Education, Abstract, PhD Thesis. http://www.sukjaro.eu/sites/default/files/JB/letoltheto/jb_kivonat_ang.pdf.

Jarosievitz B. (2011) ICT, Multimedia used in the national and international educational projects, *Informatika* **38**, pp. 22. http://www.sukjaro.eu/JBea/pdf/informatika_38_6.pdf.

Jarosievitz B. (2012) Unique activities organised for the Researchers' Night in Hungary. http://blog.scientix.eu/2015/03/16/unique-activities-organised-for-the-researchers-night-in-hungary.

Jarosievitz B. (2016) BYOD and Turn to your Neighbours, *Studies from Education and Society* Edited by Tibor János Karlovitz. International Research Institute, pp. 67-72.

Bae J.-H. & Kim S.-K. (2013) Research on Educational Use of Smart-Phone Applications with Smart Clicker Technique. In *Advances in Computer Science and its Applications. Lecture Notes in Electrical Engineering* **279** (Jeong H.S., Obaidat M., Yen N., Park J., eds.), Springer, Berlin, Heidelberg, pp. 597-602.

Juhász A. (2015) M5. Audacity akusztikus mérőprogram alkalmazása fizikaórán. *A fizika tanítása a középiskolában*, ELTE 2015, p. 702-706.

Turner K. (2015) Should we care if students enjoy lessons? https://eic.rsc.org/opinion/should-we-care-if-students-enjoy-lessons/2010071.article.

Kuhn J. & Vogt P. (2013) Applications and examples of experiments with mobile phones and smartphones in physics lessons. *Frontiers in Sensors* **1**(4), p. 67-73.

Lestik M. & Plous S. (2012) *Jigsaw Classroom*, Retrieved October 24, 2012. https://en.wikipedia.org/wiki/Jigsaw_(teaching_technique).

Lyman F. (1981) The Responsive Classroom Discussion: *The Inclusion of All Students*. In Mainstreaming Digest (Anderson A., ed.), College Park: University of Maryland Press, pp. 109-113.

Staacks S. (2016) Phyphox – RWTH Scientists Turn Your Smartphone into a Physics Lab, the 2nd Institute of Physics of the RWTH Aachen University. https://www.rwth-aachen.de/go/id/lwwh?lidx=1#aaaaaaaaaaalwwj on 12 September 2016.

Mazur E. (1997) Peer Instruction: A User's Manual Series in Educational Innovation. *Prentice Hall, Upper Saddle River*, NJ.

Foti M.K. (2014) Mobile Learning: How Students Use Mobile Devices to Support Learning, *Journal of Literacy and Technology* **15**(3), 58-78.

OECD (2000) Measuring student knowledge and skills. The PISA 2000 assessment on reading, mathematical and scientific literacy. OECD, Paris.

OECD (2001) Learning to Change: ICT in Schools, Paris, OECD. Physics Teaching/Learning at Primary Level and Teacher Education.

OECD (2005) Are students ready for a technology-rich world? What PISA studies tell us, Paris, OECD.

Grazier R. (2016) How can we teach students not to be digitally illiterate? https://www.linkedin.com/pulse/how-can-we-teach-students-digitally-illiterate-richard-grazier.

Guido R.M. (2013) Attitude and Motivation towards Learning Physics, *International Journal of Engineering Research & Technology (IJERT)* **2**(11), 2087-2093.

Sadowska M. & Kamińska A. (2010) Problems in teaching physics in primary and secondary school, as seen by young Polish she-teachers. *Proceedings of selected papers of the GIREP – ICPE-MPTL International conference Teaching and Learning Physics today: Challenges? Benefits? Reims*, August 22-27, 2010, France, pp. 180-185.

Wexler J., Brown J., Metcalf D., Rogers D. & Wagner E. (2008) eLearning guild research 360 report: Mobile learning. Santa Rosa, CA: eLearning Guild. http://www.cedma-europe.org/newsletter%20articleseLearning%20Guild/360%20Report%20on%20Mobile%20Learning%20(Jul%2008).pdf.

Wikipedia (2017) Working hypothesis, From Wikipedia, the free encyclopaedia https://en.wikipedia.org/wiki/Working_hypothesis.

Hsu Y.-Ch. & Ching Y.-H. (2013) Mobile computer-supported collaborative learning: A review of experimental research, *British Journal of Educational Technology* **44**(5), E111-114. DOI: 10.1111/bjet.12002, 2014.

IOP Conf. Series: Journal of Physics: Conf. Series **1076** (2018) 012015 doi:10.1088/1742-6596/1076/1/012015

The first year of the "open discovery of stem laboratories" (ODL) project

Dominique Persano Adorno[1], Nicola Pizzolato[1], Claudio Fazio[1], Onofrio R. Battaglia[1], Olga Dziabenko[2], Eleftheria Tsourlidaki[3], Airina Volungevičienė[4], Vida Žvinienė[4], Juri Lõssenko[5]

[1]Physics Education Research Group, Dipartimento di Fisica e Chimica, Università di Palermo, Italia
[2]DeustoTech Learning, University of Deusto, Bilbao (Spain)
[3]Ellinogermaniki Agogi, Pallini – Athens (Greece)
[4]Lithuanian Association of Distance and E-Learning – LieDM (Lithuania)
[5]Hariduse Infotehnoloogia Sihtasutus – HITSA, Tallin (Estonia)

Abstract. The Open Discovery of STEM Laboratories (ODL) project, co-funded by the European Community Erasmus+ KA2 program for 30 months, starting from November 2015, involves five countries: Spain, Italy, Greece, Estonia and Lithuania. It aims to implement teacher collaboration in creating and using μMOOCs (very short version of MOOCs-Massive Open Online Courses) for encouraging the use of STEM (Science, Technology, Engineering and Mathematics) remote/virtual laboratories into lessons. The challenge of the project is to impact on teaching process and inspire pedagogical innovation and modernization by means of open education resources, teaching/learning tools and best practices provided by European educators via the ODL platform. The ODL project consists of different phases: the setting up of the pedagogical scenarios for the design and creation of the μMOOCs, their embedding in school environments and their dissemination to a wide EU audience. The strong points of the project rely on the opportunity for teachers of improving both digital skills and pedagogical competences, experiencing international collaborative work and having the availability of attractive open education resources in national languages, helpful to design creative lessons on STEM topics. In this contribution we present the ODL project and the results obtained in the first year of its activity. In particular, we discuss the difficulty to identify the pedagogical scenarios to be adopted to create suitable μMOOCs for physics education, able to increase student ability to solve real-life problems. The benefits of adopting inquiry-based approaches, differentiated by the amount of information and teacher guidance provided to students (confirmation, structured/guided and elicited/open inquiry), will be discussed and compared.

1. Introduction

ODL Project Motivation and Objectives

The "Open Discovery of STEM Laboratories" (ODL) project has been co-funded by the European Community Erasmus+ KA2 program (Cooperation for Innovation and the Exchange of Good Practices – Strategic Partnerships for school education) for 30 months, starting from November 2015 (Project Number: 2015-1-ES01-KA201-016090). In order to support educators to find and to organize digital resources, while designing and delivering personalized instruction in school learning environments, the Open Discovery of STEM Laboratories project aims to implement teacher collaboration in creating and using μMOOCs (very short version of MOOCs-Massive Open Online Courses) for the inclusion of STEM (Science, Technology, Engineering and Mathematics) remote/virtual laboratories in the everyday teaching practices (http://opendiscoverylabs.eu).

The report issued by the Organization for Economic Co-operation and Development (OECD) "Evolution of Student Interest in Science and Technology Studies" identifies the crucial role of positive contacts with science at an early stage in the subsequent formation of attitudes towards science (Global Science Forum 2008). Recent studies show that one of the factors that influence the increase of interest, of motivation and of positive attitude towards the study of sciences generally, and of Physics in particular, is represented by the didactic methods used within the teaching-learning process. More than that, the attention should be focused on the teaching method during secondary school, which is the most important level in determining whether students prefer Science studies, since it is at this stage that they can start choosing which subjects they wish to study (Dinescu *et al.* 2011).

School atmosphere changed a lot in last decades. School education started introducing new technology, such as tablets, 3D printers, interactive whiteboards, apps and scaffolds, and other various ICT education instruments. The new learning processes should be adapted to complexity, connectivity, and velocity of new knowledge society. New curricula should be flexible to the place and time of learners, incorporate and discover the potential of new technology, and empower students to take control of their learning, to grow and move onwards. Schools should provide the "new generation" students with such learning experience which would open the doors to the best academic achievement, would ensure economic growth and civic engagement.

Laboratory teaching is an indispensable part of science education. The processes of making observations, performing systematic and quantitative investigations, data collection, analysis and logical interpretation of results and drawing relevant conclusions, are fundamental skills to the training of all science subjects. Performing experiments also serve to reinforce students' classroom learning experiences. Unfortunately, many of the scientific ideas are taught with only very limited support of the corresponding experiments, for a number of reasons. These could be cost, space and safety implications for implementing the experiments (Persano Adorno *et al.* 2015). At this respect, traditional face-to-face lectures and experimental laboratory sessions can be complemented with new online experimental frameworks (Gröber *et al.* 2007, 2008). In fact, while there already are lots of Internet resources (many of them accessible for free) to fulfil many theoretical aspects on education, engineering and scientific studies also need more specific Internet based tools to cover the practical part of their teaching. The challenge which should be met by educators as early as possible is that the inclusion of these laboratories in the curriculum is done within the frame of strategies that add value to teaching processes, giving real chances for the building of learning experiences (Concari & Marchisio 2013). Teachers should use innovative practices in their teaching through personalized learning approach and develop student critical thinking. Old traditional methods of instruction, almost exclusively based on lectures aimed at transmitting theoretical concepts to the students, are gradually replaced by methods taking into account the practice of experiments. In particular, by adopting a more constructivist view of science learning, teachers should be more oriented to stimulate the students to directly experience the natural phenomena under investigation.

In order to help schools to transform their curriculum emphasizing an academic excellence, and to support educators to employ new technological tools, creating innovative STEM school curricula, the ODL consortium offers to teachers the innovative approach based on school μMOOC development. The main innovation of this methodology is the inclusion of remote and/or virtual laboratories, but it is also ground on the developing and re-using of open education resources (OERs), the sharing of learning resources and experiences. For this purposes the project has the following objectives:

- develop a μMOOC methodology for school curriculum;
- establish the μMOOC platform adapted for STEM curriculum designing;
- train at least 300 school teachers to develop μMOOCs for STEM education by multiplier events and teacher schools;

IOP Conf. Series: Journal of Physics: Conf. Series **1076** (2018) 012015 doi:10.1088/1742-6596/1076/1/012015

- create a depository of μMOOCs for STEM education, complemented by remote and virtual laboratories, available for secondary school teachers and students.

After all collected materials will be available on the ODL platform the teachers would be able to: (i) compile the provided educational blocks (such as videos, exercises, tests, worksheets, material rationale, etc.) to create and share their own μMOOCs; (ii) merge different μMOOCs to design personalized teaching/learning paths; (iii) cooperate in European context by exchanging materials and feedbacks on μMOOC-based teaching experiences.

Fig. 1. The ODL logo

The ODL Partners

The ODL project involves five partners from different European countries: (1) DeustoTech Learning, Deusto Foundation, Bilbao-Spain; (2) Physics Education Research Group, Department of Physics and Chemistry, University of Palermo-Italy; (3) Ellinogermaniki Agogi, Pallini, Athens – Greece; (4) Hariduse Infotehnoloogia Sihtasutus – HITSA, Tallin – Estonia; Lithuanian Association of Distance and E-Learning, LieDM, Kaunas – Lithuania.

2. The first milestones of the ODL project

During the first few months of the project: (i) a close collaboration between partners has been established, (ii) local, national and transnational outputs have been discussed, (iii) decisions on quality assurance, evaluation, dissemination, exploitation, and communication plan, administrative issues-reporting, agreements, content and timetable template, property rights, financial rules and policy, working methods and participation in activities and outputs have been settled up. Moreover, the consortium partners deliberated that the first year of the project would be addressed on the finalizing of the μMOOC methodology for school curriculum and on the specification and design of the μMOOC project platform (including operational modules such as scenarios container, catalogue of STEM laboratories, teaching and learning zone, etc.).

The majority of MOOCs available today are university courses that have been put online, or courses which were created by corporations and target specific career skills. EdX platform very recently have proposed specially designed courses from top high schools, secondary schools and universities to help to prepare for Advanced Placement (AP) Exams and CLEP Exams. These courses covering subjects ranging from English language and composition to calculus, biology, statistics and computer science, give students around the world the opportunity to access quality courses and materials regardless of financial resources. Up now, to the best of our knowledge, there is no a widely accepted methodology for the realization of μMOOC to secondary school level, despite of several reports addressing the use and the efficacy of MOOC at university level (Crow 2013, Kellogg 2013, Bates 2014, Israel 2015). The greatest difficulties are related to the identification of the pedagogical scenarios to be adopted to create suitable μMOOCs for secondary school STEM curriculum, including characteristics, approaches, motivations and challenges of school teachers and to the choice of the possible frameworks and teaching strategies for the implementation in classroom, in particular in physics education.

The choice of the μMOOC Methodology for School Curriculum

To effectively determine a methodology for the design, the realization and the implementation of μMOOC at secondary school level all consortium partners decided to collect teachers' suggestions and feedback. After long debates and discussions, secondary school teachers involved in the project suggested the use of an Inquiry learning structured approach – flexible, but not too much – having different levels of complexity (Herron 1971, McDermott 1996, National Research Council 2000, Llewellyn 2002, Banchi & Bell 2008). This implies that the teaching strategies involved in μMOOC implementation should be grounded on the viewpoint that students are active thinkers, who construct their own understanding from interactions with phenomena, the environment, and other individuals. In fact, in inquiry-based learning, the students are engaged in identifying scientifically oriented questions, planning investigations, collecting data and evidences in laboratory and/or real life situations, building descriptions and explanation models, sharing their findings and eventually addressing new questions that arise.

All possible pedagogical frameworks (*scenarios*) for μMOOCs should be laboratory-based experiences and have a high degree of interactivity. Their duration will be approximately limited to 40-50 minutes, inclusive of the exploitation of remote/virtual labs. One of the keys role in making successful the proposed teaching strategy is played by the choice of the topic and learning environment. In fact, what piques student's curiosity will depend on the student's interests, experience and prior knowledge. A "good" μMOOC topic for the implementation in a classroom, should:

1. provide affective engagement to the students;
2. generate curiosity and leads to questions;
3. generate a cognitive conflict;
4. be scientifically investigated and explained within the competence of the students involved;
5. create scientific knowledge;
6. require the students to use inquiry skills to explain the involved phenomena;
7. be transversal and connected very closely to real life.

Another crucial aspect is that the laboratory has not to be considered the place where students only observe experiences carried out by others or attend fruitless demonstrations of the validity of laws previously introduced by the teacher theoretically. The students should be personally involved in experimental activities, facing problematic situations that requires reasoning efforts, in order to be solved effectively. Moreover, the laboratory activity cannot be limited to the conduction of experiments and observations, but it should include a preliminary phase characterized by posing scientifically relevant questions, designing procedures and a final critical evaluation of obtained results. Furthermore, the designing of effective μMOOC-based learning paths also include the sharing of ideas with peers, drawing explicatory models, supporting conclusions and making choices based on arguments and evidences.

The pedagogical approach to be used in the μMOOC design should take into account the 5E learning cycle (ENGAGE, EXPLORE, EXPLAIN, EXTEND, EVALUATE) to develop student critical thinking and to help students to explore and evaluate their learning (BSCS, 1993).

In particular:

(i) **Engage** state involves the setting of the learning environment in a way that piques student interest and generates curiosity in the topic under study. It get students personally involved in the lesson, while pre-assessing prior understanding. During the ENGAGE stage, students first encounter and identify the instructional task, make connections between past and present learning experiences, setting the organizational ground work for upcoming activities. The video format should arouse students' curiosity and encourage them to ask their own questions;

(ii) In the **Exploration** stage, by means of the remote/virtual labs, the students have the opportunity to get directly involved with phenomena and materials. The teacher acts as a facilitator, providing materials and guiding the students' focus. **Explore** is the beginning of student involvement in inquiry. They search for information, raise questions, develop hypotheses to test, collect data;

(iii) **Explanation** involves the process of data acquisition and evidence processing techniques for the individual groups or entire class (depending on the nature of investigation) from the information collected during the exploration. **Explain** is the stage at which students build models (descriptive or explicative), discuss their data with peers and the teacher and begin to communicate what they have learned;

(iv) **Extend** is the stage in which students expand on the concepts they have learned, make connections to other related concepts, and apply their understandings to the world around them in new ways, building possible generalizations;

(v) **Evaluate** is an on-going diagnostic process for both students and teachers. It involves students' capacity to make judgments, analyses, and evaluations of their work, also in comparison with the work of their colleagues. It also allows teachers to determine how much learning and understanding has taken place.

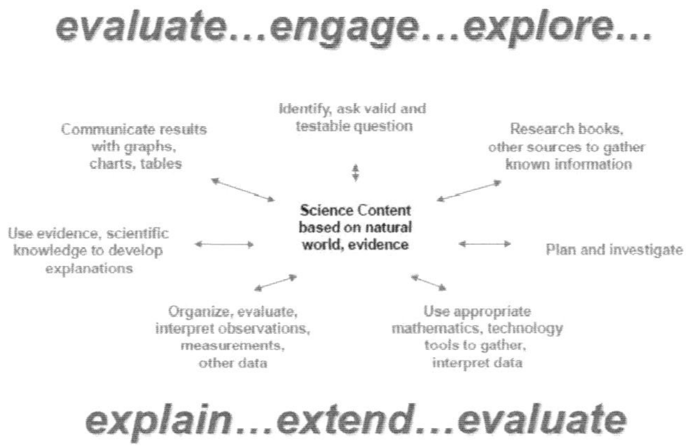

Fig. 2. The Scientific Inquiry/5E Learning Cycle

On the basis of teachers' suggestions, to ensure pedagogical coherence of all μMOOCs and, at the same time, to leave the teachers free to design personalized teaching paths, we have chosen to adopt inquiry-based approaches and differentiated the templates for μMOOC development for (i) the level of teacher guidance; (ii) the difficulty of the involved remote/virtual laboratories; (iii) the requested student's cognitive skills. In fact, there are various levels of inquiry in science education – the initial level where the teacher directs every aspect, to the highest level where the student holds the control and needs the intellectual and practical skills to become investigator, acting as a researcher. Through the series of different levels of inquiry, the student becomes more able to carry out his/her own independent inquiry, and the assistance of the teacher becomes different, less instructive, but more enabling and flexible (Pedaste *et al.* 2015).

3. Scenario development: the 3-levels inquiry-based approach

In order to develop scientific knowledge and stimulate the strengthening of reasoning skills, the students will be engaged into inquiry-based learning. The μMOOCs will be based on three key inquiry practices: 1) Coordinating hypothesis, observation and evidence; 2) Controlling variables; and 3) Studying cause and effect relationships (Kuhn 2005).

All phases of the 5E learning process will be included into the μMOOCs, but with different amount of support provided by the teacher (Zhang & Quintana 2012, Zacharia *et al.* 2015). Moreover, they should be well separated within the μMOOC in such a way that their administration could also be delayed in time.

In particular, we have chosen 3-levels Inquiry-based approach:

- *Basic* – Confirmation Inquiry

- *Intermediate* – Structured/Guided inquiry
- *Advanced* – Elicited/Open Inquiry

In order to facilitate the teachers in the transition from scenarios to educational resources we drew up the μMOOCs scenarios for the 3-levels Inquiry, by sharing key-aspects on the ODL European platform (see moocspace.odl.deusto.es).

Scenario 1: ***Basic approach*** (Confirmation Inquiry)
In the basic approach, the teacher provides students with the question, shows the use of the remote/virtual lab, illustrates the procedure and the method, but the results and their explanation are known in advance (Banchi & Bell 2008). Confirmation Inquiry is useful when the teacher purpose is to reinforce a previously introduced idea, introduce students to the experience of conducting investigations, or have students practice a specific inquiry skill, such as the collecting and recording of data.

Therefore, in this case the μMOOC topic will be previously introduced by the teacher and explained in depth; the novelty will be represented by its contextualization in real-life environments (Engage). The virtual or remote laboratories will be exploited by the teacher (Exploration). All other phases (Explanation, Extension, Evaluation) are faced and discussed by the teacher in the μMOOC. After the μMOOC vision, the students have the possibility to explore the remote/virtual experiments in class (in small groups working with tablets connected to the internet), or at home. They will be invited to write a scientific report on the experience done and on acquired concepts.

Main student outcomes: Practical applications of the theory.

Scenario 2: ***Intermediate approach*** (Structured/Guided Inquiry)
In the structured inquiry level, the question and the detailed procedure for the utilization of the remote/virtual lab are provided by the teacher. However, the students generate an explanation supported by the evidence they have collected by experiencing the remote/virtual lab by themselves. They are responsible for uncovering the answer. The teacher acts as a knowledge facilitator, providing support or materials in the μMOOC so that the students can experience a sense of success when working at this level (Banchi & Bell 2008).

Also in this case the μMOOC topic will be previously introduced by the teacher. Because this kind of inquiry is more involving than the first level, it is most successful when students have numerous opportunities to learn and practice different ways to plan experiments and record data. Therefore, after the μMOOC vision, the students should have the possibility to repeat the experiments (in class or at home) by changing the parameters. They will be invited to write a scientific report on the experience done and on acquired concepts.

Main student outcomes: Practical applications of the theory; reasoning efforts to generate explanations on the basis of their own investigation results.

Scenario 3: ***Advanced approach*** (Elicited/Open Inquiry)
In the Open inquiry the teacher takes the delicate role of defining the context for inquiry by presenting a multidisciplinary view of a theoretical problem or a real-life phenomenon. Subsequently, he/she stimulates the students to define their relevant questions, design and carry out their independent investigations, construct coherent explanations, communicate and share their results (Banchi & Bell 2008). An open inquiry-based instruction seems more efficient to reinforce learners' reasoning skills, also increasing the awareness of the process of scientific inquiry (Pizzolato *et al.* 2014). Despite this, students involved in open inquiry may develop feelings of frustration due to the lack of achieving the desired goals independently from teacher's hints (Quintana 2005).

In the Elicited/Open inquiry level, within the μMOOC the teacher will provide the students with only the research question, stimulating the learners to explore the potentialities of the remote/virtual lab by themselves. Here, the students design the procedure (method) to be followed in the use of the remote/virtual labs, record and interpret data, test their questions and share the findings. Although teachers are less instructive, they provide a framework (scaffolding) for the process when needed, prepare resource lists or support cards in order to help students to manage this level of inquiry. The students will be involved by mean of the μMOOC in a learning path with a specific process of activation – Elicited Inquiry – consisting of a learning environment in which the instructor actively will participate to the debate on the physics governing the observed experimental findings, never providing exhaustive explanations to the students, but giving comments and hints, sometimes expressly incorrect, always leaving the students in a state of uncertainty, stimulating their reasoning and activating their scientific inquiry (Persano Adorno & Pizzolato 2015).

Main student outcomes: Through self-designed or stimulated exploration students make hypotheses, test their own predictions, and draw their own conclusions; they should reach higher levels of autonomy and develop higher-order thinking skills.

4. Discussion and conclusion

The ODL project grounds on student's active learning through inquiry-based science instruction and exploitation of remote/virtual STEM laboratories. ODL methodology incorporates four key innovations: first, the use of micro-MOOCs and ICT-based educational instruments to motivate teachers in the creation of flexible personalized teaching/learning paths and to increase students' interest and involvement, due to the innovative methodology; second, the incorporation of remote/virtual laboratories, as didactical instrument for practice-based learning, with the aim to capture the students' imagination and motivation, effectively engaging them; third, the 5E cycle to develop student critical thinking and to help students to explore and evaluate their learning; and finally, the inclusion of practical exercises, evaluation tests, etc, with which students will take control and awareness of their learning process.

In this paper we present the *Open Discovery of STEM Laboratories* European Project and share the results obtained in the first year of activity. In particular, we discuss the pedagogical scenarios to be adopted to create μMOOCs, able to increase high-school student abilities and skills and give some suggestions that can guide teachers in selecting topics and useful virtual/remote labs to develop suitable μMOOC-based teaching paths promoting inquiry learning. In fact, an inquiry-based teaching environment is today considered the natural framework where to develop opportunities for learning science in terms of an active construction of meaningful knowledge. Moreover, the dealing with remote/virtual labs allows new ways of experimentation and, in educational terms, it focuses on conceptual understanding. Student's attention is focused mainly on the analysis of results that come from a real experimentation. These labs can extend the capability of a conventional laboratory and increase the number of times and places a student can perform the experiments.

The inclusion of these μMOOC-based teaching paths in the curriculum should be framed within proposals that add value to teaching, giving real learning opportunities. Our future research activity will be focused on the analysis of educational experiences carried out by our teachers in the classroom environment. Also we will investigate the evaluation criteria, from a didactic point of view, of these teaching/learning paths, in order to get information that will allow us to optimize its use for educational purposes.

References

Banchi H. & Bell R. (2008) The Many Levels of Inquiry. *Science and Children* **46**(2), 26-29.
Bates T. (2014) MOOCs: getting to know you better. *Distance Education* **35**(2), 145-148.
Biological Sciences Curriculum Study (BSCS) (1993) *Developing Biological Literacy*, Colorado Springs, CO.
Concari S. & Marchisio S. (2013) The Remote Laboratory as a Teaching Resource in the Scientific and Technological Training. *Creative Education* **4**, 33-39. DOI: 10.4236/ce.2013.410A006.
Crow M. (2013) Digital learning: Look, then leap. *Nature* **499**, 275-277.

Dinescu L., Miron C. & Barna E.S. (2011) New trends: promotion of didactic methods that favour the increase of student's interest and motivation for studying Physics. *Romanian Reports in Physics* **63**(2), 557-566.

Global Science Forum (2008) *Encouraging student interest in science and technology studies*, OECD **65**, 98.

Gröber S., Vetter M., Eckerta B. & Jodl H.J. (2007) Experimenting from a distance. *European Journal of Physics* **28**(3), 127-141.

Gröber S., Vetter M., Eckerta B. & Jodl H.J. (2008) Remotely controlled laboratories: Aims, examples, and experience. *American Journal of Physics* **76**(4), 374.

Herron M.D. (1971) The nature of scientific enquiry, *School Review* **79**, 171.

Israel M.J. (2015) Effectiveness of Integrating MOOCs in Traditional Classrooms for Undergraduate Students. *The International Review of Research in Open and Distance Learning* **16**(5), 102-118.

Kellogg S. (2013) Online learning: How to make a MOOC. *Nature* **499**, 369-371.

Kuhn D. (2005) *Education for thinking*. Cambridge, MA: Harvard University Press.

Llewellyn D. (2002) *Inquiry Within: Implementing Inquiry-based Science Standards*, Corwin Press Inc., Thousand Oaks, CA.

McDermott L.C. & Physics Education Group at the University of Washington (1996) *Physics by inquiry, Volumes I and II.* New York: Wiley.

National Research Council (NRC) (2000) *Inquiry and the National Science Education Standards: A Guide for Teaching and Learning*, The National Academies Press, Washington DC.

Pedaste M., Mäeots M., Siiman L.A., de Jong T., van Riesen S.A.N., Kamp E.T., Manoli C.C., Zachariac Z.C. & Tsourlidaki E. (2015) Phases of inquiry-based learning: Definitions and inquiry cycle. *Educational Research Review* **14**, 47-61.

Persano Adorno D. & Pizzolato N. (2015) An inquiry-based approach to the Franck-Hertz experiment, *Il Nuovo Cimento* C **38**, 109. DOI: 10.1393/ncc/i2015-15109-y.

Persano Adorno D., Pizzolato N. & Fazio C. (2015) Elucidating the electron transport in semiconductors via Monte Carlo simulations: an inquiry-driven learning path for engineering undergraduates. *European Journal of Physics* **36**(5), 1-19.

Pizzolato N., Fazio C., Sperandeo Mineo R.M. & Persano Adorno D. (2014) Open-inquiry driven overcoming of epistemological difficulties in engineering undergraduates: A case study in the context of thermal science. *Physical Review ST Physics Education Research* **10** 010107. DOI: 10.1103/PhysRev STPER.10.010107.

Quintana C., Zhang X. & Krajcik J. (2005) A framework for supporting metacognitive aspects of online inquiry through software-based scaffolding. *Educational Psychologist* **40**(4), 235-244.

Zacharia Z.C., Manoli C., Xenofontos N., de Jong T., Pedaste M., van Riesen S.A.N., Kamp E.T., Mäeots M., Siiman L. & Tsourlidaki E. (2015) Identifying potential types of guidance for supporting student inquiry when using virtual and remote labs in science: A literature review. *Educational Technology Research and Development* **63**, 257-302.

Zhang X. & Quintana C. (2012) Scaffolding strategies for supporting middle school students' online inquiry processes. *Computers & Education* **58**(1), 181-196.

IOP Conf. Series: Journal of Physics: Conf. Series **1076** (2018) 012016 doi:10.1088/1742-6596/1076/1/012016

PCK for introductory mechanics pre- service teachers in a conceptual lab

Brigitte Wolny, Martin Hopf

University of Vienna, Austrian Educational Competence Centre Physics

Abstract. The Physics School Experiments Laboratory (PSE-lab) lab at the University of Vienna aims to enhance pre-service teachers' pedagogical content knowledge (PCK). In this paper we present a design- based research (DBR)-project which we use to develop and evaluate the mechanics section of this lab course. Attendees of the PSE-lab are pre-service physics teachers in the 4th semester of their studies for a Bachelor of Education. Within this lab course pre-service physics teachers learn introductory mechanics with a non-traditional teaching-learning sequence, starting with two-dimensional dynamics. The present project keeps track of pre-service teachers' conceptions and explores their growth in terms of PCK during four units, 90 minutes each, of PSE-lab. The hypothesis is that, while following the new content structure for introductory mechanics, pre-service teachers will develop a deeper understanding of Newton's laws on the one hand and on the other hand they will gain insights on how to implement such a structure into their own future teaching. Qualitative analysis of pre-service teachers' interviews is used to characterize pathways of PCK-development. Despite this short intervention of six hours we have seen clear evidence in pre-service teachers' early PCK development.

1. Introduction

Pedagogical content knowledge (PCK) has received much attention in recent years due to teachers' impact on students' learning (Lipowsky 2006). By putting more emphases on pre-service teachers and their courses at university level, the opportunity is given to increase teachers' PCK before starting as practitioners. However, there are few research projects describing early stages of pre-service teachers' developing PCK. Although pre-service teachers' PCK is assessed professionally with tests and inventories, learning opportunities during their university studies are rarely documented in detail. The present paper describes the development of such a learning opportunity for pre-service teachers within the Physics School Experiments lab course (PSE-lab).

2. Theoretical background

Pre-service physics teachers are to learn not only physics and pedagogy, but also develop a variety of skills and aspects of pedagogical content knowledge (PCK). "Teaching is demanding and difficult mental and physical work that only the most well- educated and mentored professionals can accomplish. PCK is an attribute that teachers develop, and it cannot be found among mere subject matter experts or among those who are 'good with kids'" (Shulman 2015). Nearly 30 years ago Shulman conceptualized a teacher's knowledge base as complex and broad-ranging, comprised of subject matter knowledge, curricular knowledge and subject matter for teaching, consisting of seven distinct bases (Shulman 1987, p. 8). Numerous further conceptualizations have been found in the meantime as Gess-Newsome (2001) describes systematically in her book. Implications for science education are discussed (Gess-Newsome & Lederman 2001), at the same time as conceptualization of teacher's knowledge bases went on. Lee (2007) similarly found seven categories through which experienced teachers' PCK could be described. Researchers are convinced that PCK is not only experimental knowledge acquired through classroom teaching but also an "integrated set of knowledge, conceptions, beliefs, values which teachers develop in the context of teaching situation" (Lee *et al.* 2007). The present project deals with pre-service physics teachers and not with practitioners. We conceptualized a pre-service teacher's knowledge base for physics teaching comprising six main parts (see Krumphals 2017):

1. General knowledge of learning processes (e.g. motivation, context, nature of science);
2. Knowledge about evaluating learning processes (e.g. cognitive levels of pupils' answers, diagnosis of learning difficulties);
3. Knowledge about how to plan learning processes (e.g. choosing adequate content);
4. Knowledge about the general framework (e.g. goals of physics learning, curriculum);
5. Knowledge of pupils' learning processes (e.g. pupils' beliefs, content specific learning difficulties, motivational factors);
6. Knowledge about implementing experiments (e.g. role of class room experiments, demonstration and learning effectiveness of experiments).

Within the PSE-lab we want to facilitate only two bases, number three and number six in the latter numeration. Pre-service physics teachers are to learn about how to plan learning processes and they shall gain knowledge about implementing experiments in such a way as to enhance their future middle school students' learning.

Educational Reconstruction for Teacher Education

Pre-service teachers retain their own beliefs about their future students' preconceptions, representations of subject matter and perception of teaching. Van Dijk presented a research model to study science teachers' PCK (Van Dijk & Kattmann 2007). Within her Educational Reconstruction for Teacher Education (ERTE) framework, teachers' knowledge is related to developing learning activities for students (pupils) which includes the design of teaching-learning sequences, the study of students' pre-scientific conceptions, and subject matter analysis.

These three main parts are all subject to research: 1) the learner's perspective, 2) the content to be learned, and 3) the learning environment itself. Our idea is the development of a learning environment on the bases of learners' perspective (beliefs and difficulties with a specific physics content) and expert's opinion on subject matter (content) which should be learnt. At the University of Vienna, we didn't develop a learning environment for pupils, but on a higher level, we started designing a university course for pre-service teachers, the Conceptual Lab in Introductory Mechanics.

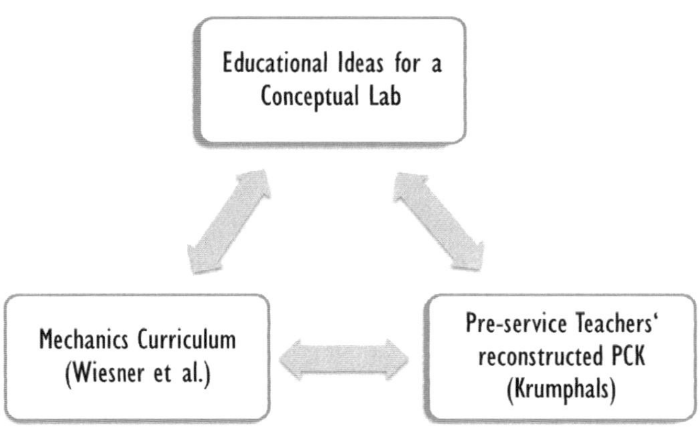

Fig. 1. Educational Reconstruction of pre-service physics teachers' Conceptual Lab in Introductory Mechanics

Fig. 1 shows on the learners' side (right), pre-service teachers' beliefs about teaching and pedagogical content knowledge. This first edge of the triangle was investigated by Krumphals (2017) in her Ph.D. studies. On the content side (left), there is a Mechanics Curriculum (Wiesner *et al.*) for pupils, which was evaluated in Germany (Tobias, 2010). Future teachers shall know about empiric evaluated curricula and the specific structure of content matter, this is seen as one part of their PCK. Therefore this Mechanics Curriculum for pupils, represented in a booklet (Hopf *et al.* 2011), was elected to represent the learning content for physics pre-service teachers. On the top, the third edge of the ERTE framework triangle, there is the developing learning environment, which is represented by the Physics School Experiments laboratory and especially four units called the Conceptual Lab in Introductory Mechanics. Our hypothesis is that, while following the new content structure for introductory mechanics with this booklet, physics pre-service teachers will develop a deeper understanding of Newton's laws on the one hand and they will gain insights on how to implement such a structure into their own future teaching on the other hand. We investigate the influence of this new learning opportunity on pre-service teachers' PCK.

3. Method

Design Based Research

The present design based research project (DBR Collective, 2003) focuses on pre-service teachers' development of PCK and takes their conceptual understanding in mechanics into account. This project keeps track of pre-service teachers' conceptions (written protocols, Force Concept Inventory (FCI)) and explores pre-service teachers' gain on PCK (interviews) during an six hour teaching-learning sequence in introductory mechanics (four units of Introductory Mechanics).

Following this method several cycles of design, enactment and study were undertaken in order to find proto-theories of learning.

Initial research plan

Using the framework of ERTE we related research from our group about pre-service physics teachers' ideas and beliefs during their studies at University of Vienna (Krumphals & Hopf 2012) to a Mechanics Curriculum (Hopf *et al.* 2010). Built upon these results a new teaching-learning sequence for pre-service teachers was implemented in the conceptual lab (see Fig. 1). We intend to describe parts of pre-service teachers' PCK and their learning strategies in the conceptual lab.

Conceptual Lab for Physics School Experiments

Physics School Experiments (PSE) lab encloses a number of units and topics on teaching physics at secondary school. It aims at enhancing pre-service teachers' PCK as getting conceptual understanding in mechanics especially for school classes. Pre-service teachers will get an experimental perspective on physics concepts and come to recognize that experiments performed in the classroom generally have a different purpose than those performed in conducting scientific research. Pre-service teachers are expected to be familiar with the national high school curriculum for physics. Prior to the lab session they have to select relevant school experiments. In the lab, these experiments are conducted with typical school apparatus. During the lab, pre-service teachers reflect together with supervisors on the function of their selected experiments for middle and high school physics teaching. At the same time teaching goals and the Austrian science education standards play a major role. A deeper reflection on learning efficacy of school experiments is done in a seminar afterwards. Fig. 2 gives an overview about each unit of the PSE lab and summarizes crucial goals of each part of the lab like, preparation work, experimental work and writing exercises.

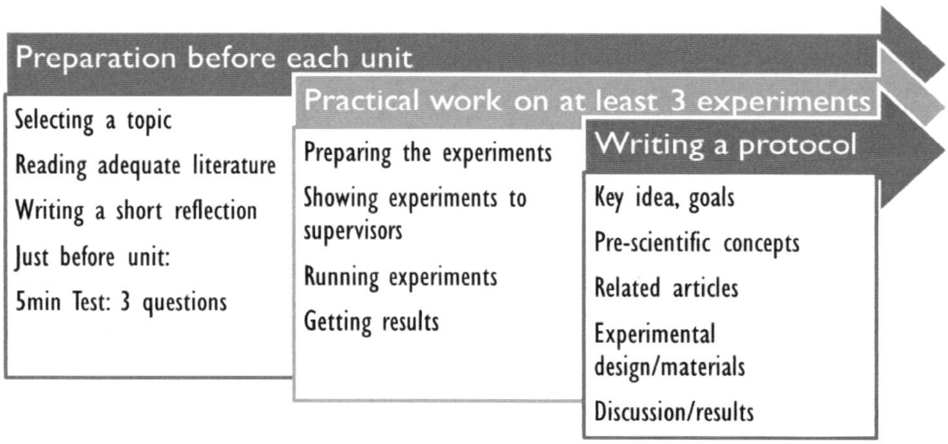

Fig. 2. Design of each unit within the Physics School Experiments lab. The term "pre-scientific concepts" means, that pre-service teachers describe misconceptions their future students will likely hold

Introductory Mechanics – Description of the Design

At the University of Vienna pre-service physics teachers attend the Physics School Experiments Laboratory, PSE-lab in the 4th semester of their Bachelor studies. During this PSE-lab, they engage themselves with a non-traditional teaching-learning sequence in introductory mechanics (Hopf *et al.* 2010). This alternative content structure includes the key idea of starting with two-dimensional dynamics, for a better understanding of Newtonian mechanics (Tobias 2010).

Four units, 90 minutes each, are to be attended chronologically:

- M1 Description of motion
- M2 Newton's first law
- M3 Newton's second law
- M4 Newton's third law

Pre-service teachers use a mechanics booklet (Hopf *et al.* 2011), initially designed for (10-12 year old) students in introductory mechanics.

In Design Cycle 2 a compulsory activity was added: Three writing activities to reflect on the mechanics curriculum and the material (Mechanics Booklet).

Lab journal entries, so-called protocols are written after each unit (M1-M4) to prepare for future experimental classroom activities.

Goals within the four units of PSE Lab

The main goals are that pre-service teachers learn about the content structure of this teaching-learning sequence as well as knowing how to implement experiments to enhance their students' learning. Our hypothesis is that, while strictly following the new content structure for introductory mechanics, pre-service teachers will develop a deeper understanding of Newton's laws as well as gain a better perception of how to implement such a structure into their own teaching. In short, we expect that learning to use a proven teaching experiment ('recipe') will help the pre-service teachers to gain PCK effectively.

Design Cycles

As the project is a DBR project, it aims at gaining proto-theories of learning. Our question is: *How do pre-service physics teachers at the University of Vienna acquire PCK in PSE lab?*

Continuous cycles of design, implementation and analysis are needed to give an answer to this question (see Fig. 3). After each cycle relevant implications for teaching in the PSE lab are communicated with instructors and feedback is given by pre-service teachers. This process allows an evaluation of the design in this authentic framework of a lab.

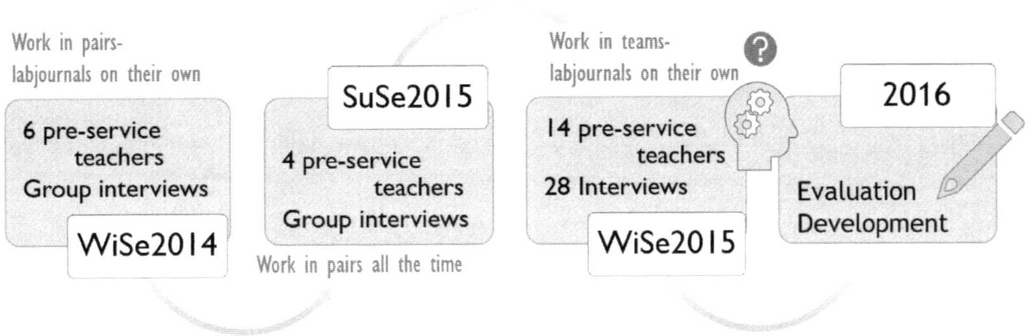

Fig. 3. Timeline of DBR Cycles: winter semester (WiSe) 2014, summer semester (SuSe) 2015 and winter semester 2015. Numbers represent pre-service teachers which were interviewed, usually there are around 20 pre-service teachers in each lab course

Design Cycle 1.
In winter semester 2014 first cycle started with one main research question: *How do pre-service teachers work with the Mechanics Booklet?* Experiments were documented by each pre-service teacher in a laboratory journal (protocol). To assess the development of PCK, the author 1) analyzed these journals, 2) observed the pre-service teachers in class during the four mechanics units, and 3) conducted guided interviews with three pairs of pre-service teachers before and after the lab and analyzed interviews with qualitative content analysis. The details and results of this assessment are reported in Wolny & Hopf, 2016.

Design Cycle 2.
First change was implemented in summer semester 2015, pre-service teachers handed in their protocols in pairs, having written a protocol collaboratively. Guided interviews (two groups of two pre-service teachers) before and after the lab were conducted and analyzed with qualitative content analysis. A new question arose: *What helps physics pre-service teachers to reflect more on the Mechanics Booklet?*

Design Cycle 3.
In winter semester 2015 two pairs of pre-service physics teachers worked together in teams (two pairs of pre-service teachers together) during the lab. Reflective questions about the Mechanics Booklet and about the mechanics curriculum had to be answered before attending the lab. Pre-service teachers handed in their protocols in pairs, having written a protocol collaboratively. Guided interviews (each pre-service teacher on his/ her own) before and after the lab were conducted and analyzed with qualitative content analysis.

Interview analysis
Within the interview guideline pre-service teachers were not only asked to describe a possible learning path for mechanics (concept map) but also to reflect on fictitious classroom activities. In the interview before the lab there was only one prompt (see Fig. 4 on the left.), whereas after attending the lab interviewees were given two different prompts to comment on differences.

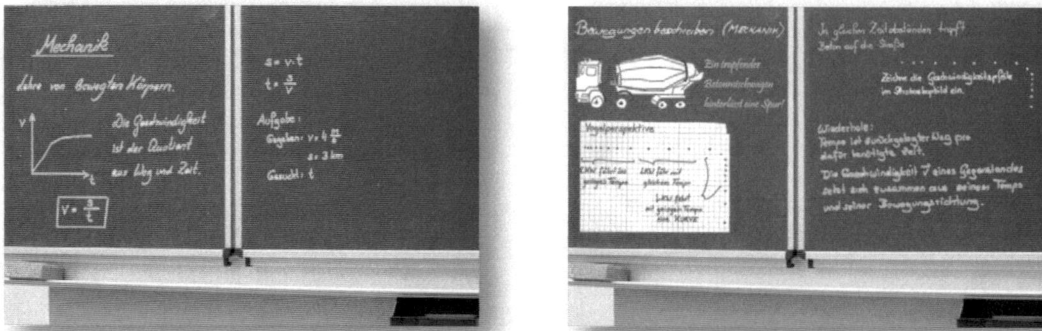

Fig. 4. Interview prompts: Blackboard-pictures of classroom activity 'introduction of velocity' On the left: a mathematical approach with formula for velocity and diagram. On the right: a strobe picture with focus on difference in position and period of time (blackboard content in German)

As not all of the pre-service teachers finished all parts of the conceptual lab (M1-M4) in sequence, and the interviews were conducted voluntarily, the number of actual analyzable interviews varied each semester. In winter semester 2014 there were 6 out of 17 pre-service teachers interviewed, in summer semester only 4. During the main study in winter semester 2015 pre and post interviews with 14 out of 22 pre-service teachers were conducted (see Fig. 3).

4. Results

The first study of pre-service teachers' developing pedagogical content knowledge was conducted in 2014 (Wolny & Hopf, 2016), followed by a redesign of the teaching-learning sequence (four units on introductory mechanics of Physics School Experiments laboratory (PSE-lab)) and a second study in 2015. Depending on pre-service teachers' *prior teaching-experiences* and *commitment during the PSE-lab*, they describe learning settings in a more professional way after the lab, using terminology, physics concepts and correct descriptions in their interviews. Knowledge about implementing and arranging learning-settings is broadened, whereas the conceptual understanding is only slightly enhanced, shown with a Hake's gain of 0.2 on FCI.

Development of PCK

In interviews before attending PSE-lab pre-service teachers recognized only a mathematical approach and the importance of language for teaching sequences. They very rarely formulated alternative learning setups, but wanted to look into the physics textbook and read the curriculum. Often interviewees mentioned acceleration and forces as concepts to start teaching introductory mechanics.

In post-interviews pre-service teachers recognized the importance of starting introductory mechanics with time and position and elucidating a concept of velocity as a term for speed and direction. Scaffolding in introductory mechanics was described with concrete examples of experimental setups. Some pre-service teachers already mentioned students' views and problems and combined teaching sequences with examples from everyday life.

The material used in the four introductory mechanics units, especially the Mechanics Booklet, is highly appreciated. Pre-service teachers like to get an insight in how implementation in teaching-sequences is arranged. It seems that they have a better guideline, time management and orientation to design experimental activities for a given context and physics topic. Design Cycle 1 showed that at this early stage of studies (4th semester of Bachelor studies) pre-service teachers weren't able to reflect on the materials independently. Design Cycle 2, underwent a smaller change due to changing instructors in the PSE-lab. Pre-service teachers were working on protocols collaboratively. This collaborative work in pairs was maintained and even elaborated in the following design cycles.

In Design Cycle 3 mandatory reading exercises and reflecting questions were added to the former design of the lab-units (Feedback from Cycle 1). Two pairs of pre-service teachers worked together on one unit as this supported discussion and reflection on possible student-prior knowledge (pupils' pre-scientific concepts) and key ideas of the curriculum (Feedback from Cycle 2). The collaborative work on each unit is favored by pre-service teachers, but the learning progress is still dependent on the individual dedication to work in the PSE-lab.

5. Discussion & conclusions

Due to changes in the University Curriculum and changing instructors in the PSE-lab, many important factors for successful learning are also under constant change. But there seem to be at least three facts supporting a positive early development of PCK.

- The Mechanics Booklet (middle school students' version) is a helpful tool for pre-service physics teachers to plan learning processes for their future students
- Collaborative work in teams (four pre-service teachers) enhances discussion and reflection on pre-scientific conceptions
- Knowledge of mechanics curriculum helps pre-service physics teachers to implement experiments in a teaching-learning sequence

Taking content knowledge into account, pre-service teachers' knowledge base at the beginning of the 4th semester varies a lot. The four units of introductory mechanics showed a gain in conceptual thinking in mechanics but an even higher impact on PCK, according to the concepts of how to plan learning processes in introductory mechanics and how to implement experiments, which supports our hypothesis.

The four units M1 to M4 should be done chronologically which is not always possible within the lab setup. We cannot make an assumption about learning outcomes in other settings so far. We observed, that leaving out the guiding set of reflection questions, leads to experimenting pre-service teachers without thinking during their lab work. While reading through protocols, supervisors can tell which pre-service teacher did prepare for reflection questions or not.

Further analysis is needed to combine findings of interview-analysis with written protocols of each pre-service physics teacher.

References

Gess-Newsome J. (2001) The professional development of science teachers for science education reform: A review of the research. *Professional development: Planning and design*, 91-100.

Gess-Newsome J. & Lederman N.G. (eds.) (2001) *Examining pedagogical content knowledge: The construct and its implications for science education* **6**. Springer Science & Business Media.

Hopf M., Tobias V., Waltner C., Wiesner H. & Wilhelm T. (2010) Influence of content structure on students' understanding of Newtonian mechanics – results from an empirical study. *GIREP Conference 2010.*

Hopf M., Wilhelm Th., Waltner C, Tobias V. & Wiesner H (2011) Einführung in die Mechanik. ('Introduction to Mechanics'). München, Würzburg, Wien, published by the authors themselves, 4th edition. Available online at https://aeccp.univie.ac.at/fileadmin/user_upload/z_didaktik/AECC_Physik/Lehrer_innen/Physik_verstaendlich/Mechanik/Mechanikbuch_Druckversion.pdf on 23 March 2018.

Krumphals I. & Hopf M. (2012) Physiklehramtsstudierende–Vorstellungen zu Studium und Physiklernen, In *2012 – Konzepte fachdidaktischer Strukturierung für den Unterricht* (Bernholt, Hg.), pp. 215-217.

Krumphals I. (2017) *Vorstellungen von Physiklehramtsstudierenden zu Physikunterricht und Studium*, PhD-Thesis, University of Vienna.

Lee E., Brown M.N., Luft J.A. & Roehrig G.H. (2007) Assessing beginning secondary science teachers' PCK: Pilot year results. *School Science and Mathematics* **107**(2), 52-60.

Lipowsky F. (2006) Auf den Lehrer kommt es an. Empirische Evidenzen für Zusammenhänge zwischen Lehrerkompetenzen, Lehrerhandeln und dem Lernen der Schüler In *Kompetenzen und Kompetenzentwicklung von Lehrerinnen und Lehrern* (Allemann-Ghionda Cristina, Hrsg.; Terhart E-wald, Hrsg.) Weinheim u.a.: Beltz 2006, pp. 47-70 (Zeitschrift für Pädagogik, Beiheft; 51).

The Design-Based Research Collective (2003) Design-based research: An emerging paradigm for educational inquiry. *Educational Researcher*, 5-8.

Tobias V. (2010) *Newton'scheu Mechanik im Anfangsunterricht, Die Wirksamkeit einer Einführung über die zweidimensionale Dynamik auf das Lehren und Lernen* **105**, Logos Verlag Berlin GmbH.

Shulman L.S. (2015) PCK: Its genesis and exodus. In *Re-examining pedagogical content knowledge in science education*, (Berry A., Friedrichsen P. & Loughran J., Eds.), Routledge, pp. 3-13.

Shulman L. (1987) Knowledge and teaching: Foundations of the new reform. *Harvard educational review* **57**(1), 1-23.

Van Dijk E.M. & Kattmann U. (2007) A research model for the study of science teachers' PCK and improving teacher education. *Teaching and Teacher Education* **23**(6), 885-897.

Wiesner H., Tobias V., Waltner C., Hopf M., Wilhelm T., & Sen A.I. (2010) Dynamik in den Mechanikunterricht. *PhyDid B-Didaktik der Physik-Beiträge zur DPG-Frühjahrstagung.*

Wolny B. & Hopf M. (2016) Einsatz des Münchner Mechanik-Konzeptes in der Lehramtsausbildung. In *Authentizität und Lernen – das Fach in der Fachdidaktik* (Maurer C., Hrsg.), Gesellschaft für Didaktik der Chemie und Physik, Jahrestagung in Berlin 2015, Universität Regensburg, p. 113.

IOP Conf. Series: Journal of Physics: Conf. Series **1076** (2018) 012017 doi:10.1088/1742-6596/1076/1/012017

Modeling mechanical, magnetic and thermal processes in high school lab activities: an experiment with a rotating disc

Tommaso Corridoni[1], Michele D'Anna[2]

[1]Liceo cantonale di Locarno and SUPSI DFA Locarno (Switzerland)
[2]Liceo cantonale di Locarno (Switzerland)

Abstract. In this paper we present the case study of a rotating disc stopped by a dissipative interaction arising from the relative motion between some magnets, fixed on top of the disc, and a copper plate kept at rest at a very short distance above them. We measure the time dependence of the angular velocity of the disc and the temperature change in the copper plate, during both the braking phase and the subsequent cooling process. The experiment is analysed and modelled step by step:

- from the *mechanical* point of view, the dry friction (always present) and the magnetic braking interaction are separately modelled: indeed, comparing the model results with the collected data, from the measured angular velocity vs. time it is possible to determine the value of all the phenomenological coefficients;
- from the *thermal* point of view, the model stresses the quantitative connection between the dissipative process (due to the eddy currents generated by the magnetic interaction), the thermal conduction with the surroundings, the temperature changes of the copper plate and that indicated by a thermometer placed inside it;
- from the *energy* point of view, testing this way the coherence of the whole model.

From a didactical perspective, each sub-model emerged step by step developing independently the various parts of the model can be considered as a quantitative conceptual map, and represents therefore a powerful tool for interaction with the thought schemes used by the students.

1. Introduction

Teaching physics in high school is often limited to situations and examples in which the interconnections between the several aspects of a real process are in fact decoupled or artificially suppressed, determining the recurrent and persistent difficulties, experienced and documented by both teachers and students in constructing a coherent and consistent teaching sequences, in particular in an interdisciplinary perspective. This is due partly to the comprehensible concern to limit the inquiry to situations in which only one aspect at a time is presented and highlighted; and partly because the more complex situations often require advanced mathematical tools for a quantitative description. This case study can help also (advanced) high school students to bridge this gap: today indeed, dynamical modelling, exploiting numerical methods well supported by easy-to-use software, provides an important tool to enable access to a large number of interesting situations even at a high school level.

This activity that represents one of the possible destinations of a teaching approach, whose general framework is discussed in a separate communication presented at this seminar (D'Anna 2016) and founded on the following points: 1) a rigorous and systematic distinction between being *constant* and being *conserved*; 2) the systematic link to the *production of entropy* for all processes in which *energy dissipation* occurs; 3) the introduction of *process diagrams* as graphic tools to allow students to follow

IOP Conf. Series: Journal of Physics: Conf. Series **1076** (2018) 012017 doi:10.1088/1742-6596/1076/1/012017

both the energy exchanges *between* a system and its surroundings, and the energy transfer (release and/or upload) *inside* a system during a process; 4) the introduction of *fields* as real physical systems that can store or transport energy.

This approach is significantly fostered by several experimental situations which can be presented to the students at the different levels both as demonstrations experiments and lab activities. Regarding the introduction of the energy concept, a series of experimental situations dealing with analogies were presented in a earlier GIREP conference (D'Anna 2010), while a detailed analysis of the energy flows and energy balance of a magnetically damped mechanical oscillator, to clarify the different roles played by the spring forces and the dissipative force(s) in the energy exchanges, was presented in (Corridoni *et al.* 2014).

Here the case study is introduced by two preparatory experimental situations that can be presented to the students as lab activities at an early stage of the teaching sequence. These experiments were also chosen in order to show how this approach to energy naturally enhances the interconnections between different aspects of natural phenomena and also allows students to be introduced to dynamical modelling activities.

2. Preparatory lab activities focused on friction and dissipation

This sequence of lab activities concerning friction processes, starting from simple mechanical observations, is aimed at testing the conservation of energy, by measuring the thermal aspects at the same time.

Sliding box. The first experiment can be carried out in classroom and/or executed directly by students in lab activities. A box of total mass *M*, containing some chalk sticks, is set in motion manually along the table, towards a motion sensor which starts measuring when the box approaches it more than a predetermined distance (Fig. 1a). The velocity of the box linearly decreases in time (Fig. 1b), so that students can conclude that the friction braking force is constant in time, determining also its value.

a)

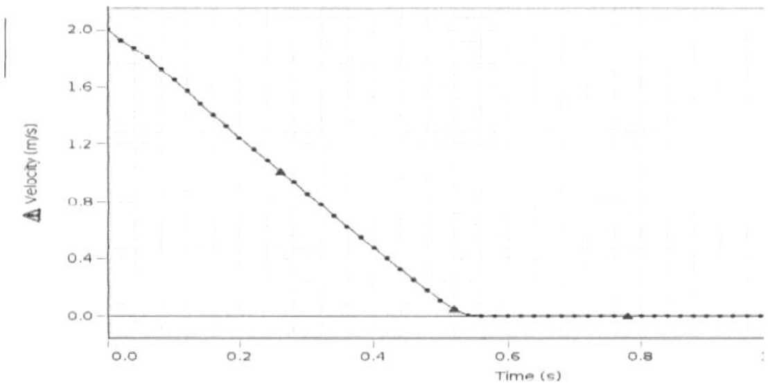

b)

Fig. 1. The sliding box experiment: (a) experimental set up; (b) box's velocity vs time

Thus they can analyse quantitatively the experimental results from the point of view of the energy exchanges, representing graphically the value of the energy flow which goes out of the system vs. time, and determining (graphically and algebraically) the amount of energy dissipated. Energy conservation can then be verified by comparing this value with the initial kinetic energy.

Magnetically braked cart. As a challenge, a slightly modified version of the previous experience can be presented to the students, with at present a time depending friction: some neodymium magnets are fixed under a cart, thrown manually along an aluminium track (Fig. 2a). The experiment is repeated under the same conditions as in the previous example, the students basically receiving the same questions. However, for students, it is a more challenging situation, because the time dependence of the velocity makes it impossible for them to reach the result algebraically (Fig. 2b).

a)

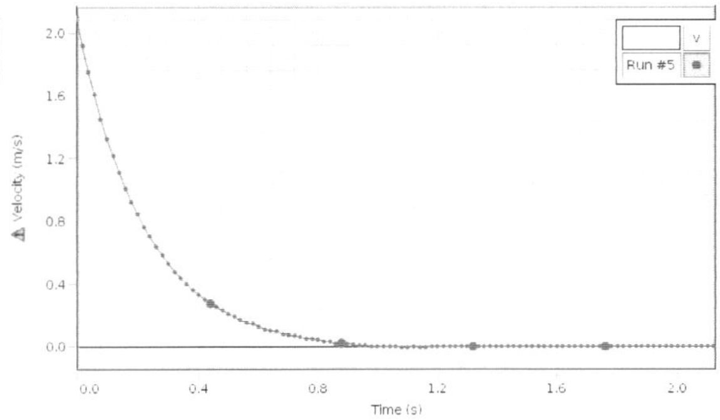

b)

Fig. 2. A magnetically braked cart: (a) experimental set up; (b) cart's velocity vs time

Students can be asked to comment on this new situation in words, to represent the exchange of energy between the cart and the track graphically, as well as to predict the value of the energy dissipated in the whole process. If the experiment is carried out with students of advanced courses[11], it is possible to use this example to introduce them to the dynamics modelling techniques, to let them grasp precisely the different role of physical relationships[12]. For example, they learn to distinguish between general relations that structure the whole physical description of natural phenomena (i.e. principles, general laws…), and the particular relationships characterizing individual systems (constitutive laws).

This version of the braked cart experiment can also be useful to address the *effects* of energy dissipation: the answer generally provided by the students – as obvious as it is problematic – is that "somewhere heat is produced", so that, in theory, they expect both the cart as well as the track to be heated. An experimental verification of this prediction (with the equipment described above) is practically impossible, since the dissipation happens all along the track. In order to measure the thermal effects it is necessary to *localize* the dissipation process, which is exactly the aim of the next experiment.

3. Localizing the dissipative process: the rotating disc experiment

A plastic disc is made to rotate by hand and rapidly braked by the dissipative interaction between some magnets, fixed on the disc, and a little copper plate (Fig. 3a), held just above the magnets and thermally insulated from the surroundings as efficiently as possible. Fig. 3b shows the detail of the thermometer inserted in the plate to monitor the thermal effects of this dissipative process. A short video of the experiment can be downloaded (QR code, Fig. 3c).

The size of the disc, the number and the strength of the magnets and the dimensions of the copper plate are chosen to allow measuring the heating of the copper plate with school equipment. Since there are many side effects occurring simultaneously, the complete dynamical model is obtained assembling the models relative to two preparatory experiments, whose parameters are determined independently comparing the measured data with models predictions[13]. From the didactical point of view, each submodel can be considered as a *quantitative conceptual map*: it represents a powerful and concrete tool to interact with the thought schemes used by the students.

---

[11] In our educational system there are optional courses that students can choose during the last two years of high school. Among these there is also *Fisica e applicazioni della matematica* (Physics and mathematics applications).

[12] For an introduction to dynamical modeling, see for example (Fuchs 2002).

[13] In particular the moment of inertia of the various parts of the rotating system (axle, disc with magnets) are experimentally determined measuring the angular velocity increase / decrease under the action of a constant torque.

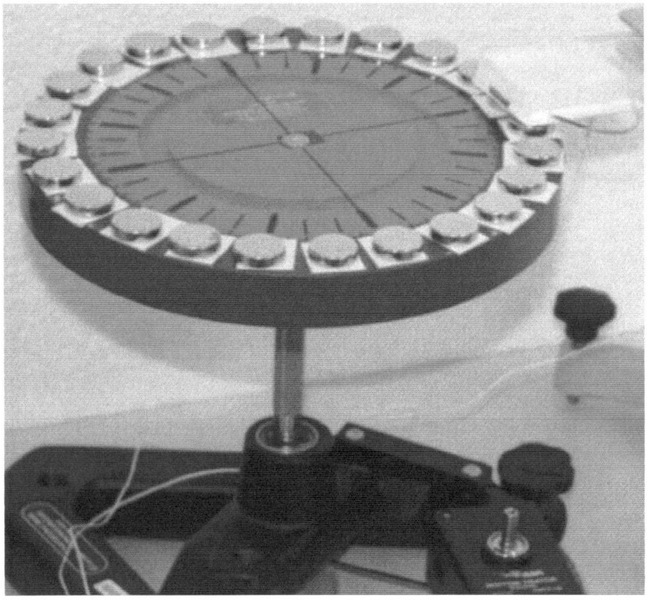

a)

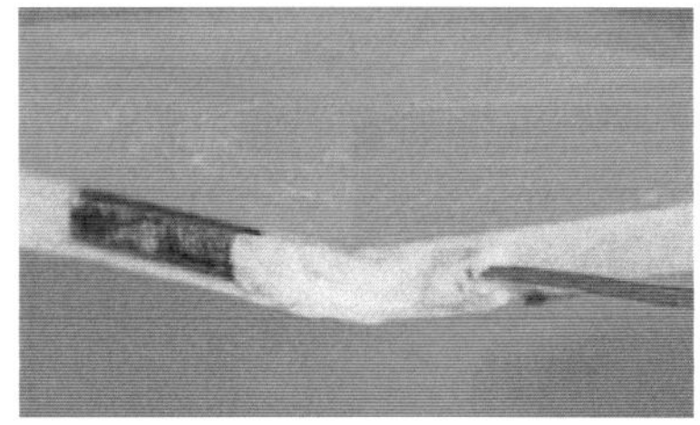

b)

c)

Fig. 3. A magnetically braked rotating disc: (a) experimental set up; (b) thermometer inserted in the copper plate; (c) QR code to download a video of the experiment

In a first experiment, the disc is put in rotation after the copper plate has been removed: it is observed that the rotation is slowed down and that the disc gradually stops. By measuring the angular velocity decrease due to the friction torque, it is possible to fix this "mechanical" parameter. Fig. 4a shows the scheme used to model the situation: the core is essentially the *balance equation* for the angular momentum. Fig. 4b shows both the model's prediction and the data measured for the angular velocity: the essentially linear decrease allows for the modelling of this interaction as dry friction.

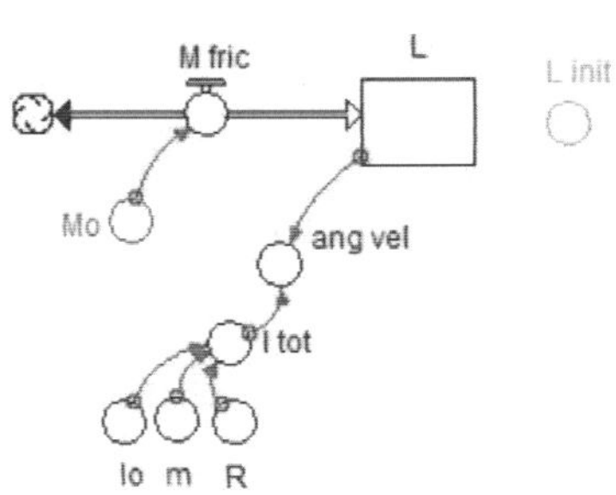

a)

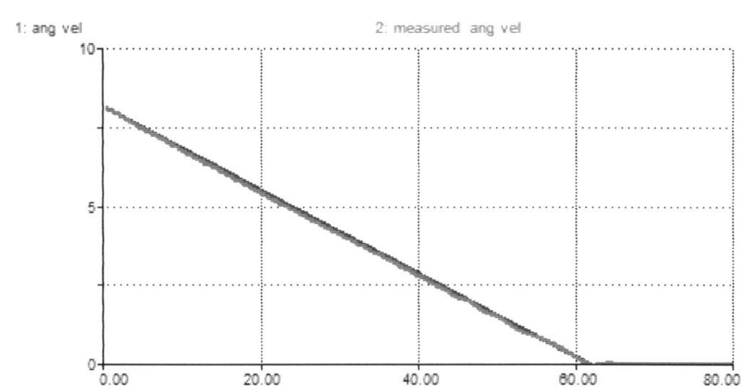

b)

Fig. 4. A rotating disk braked by dry friction only: (a) model; (b) angular velocity vs time: simulated (1 – blue) and measured (2 – red) values

In a second preparatory experiment, the temperature of the copper plate is measured in time after the disc has been magnetically braked, and consequently the copper plate heated. By measuring the temperature of the copper plate during the cooling, it is possible to fix all the thermal parameters characterizing its interaction with the surroundings (to obtain repeatable results, it is important to operate in a room which is absolutely free of air currents). Fig. 5a illustrates the corresponding model. The thermal process is modelled assuming that the copper plate is involved in heat exchanges – at different rates – with both the insulating support as well as with the thermistor acting as a thermometer. The thermal interaction with the surroundings is supposed to be only of a conductive type. The thermal capacities of the thermistor and the support as well as the different conduction coefficients are fixed by matching the model predictions with the measured data. The result is shown in Fig. 5b.

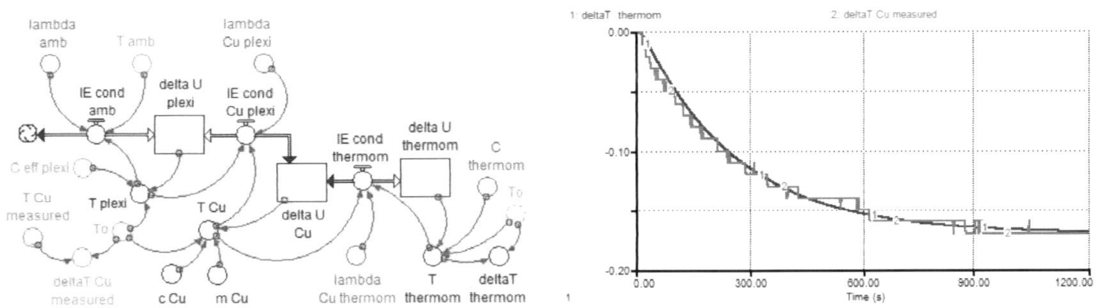

a) b)

Fig. 5. A rotating disk braked by dry and magnetic friction: (a) model; (b) temperature of the copper plate after the braking vs time: simulated (1 – blue) and measured (2 – red) values

At this point all the system parameters are set, and the experience on which we are focused can be performed: data collection is started and the disc, giving a vigorous initial spin by hand, is put into rotary motion. This time the disc requires a much shorter time interval to stop: this is due to the action of the magnetic brake. The trend registered for the angular velocity (Fig. 6a) suggests that a braking force of a viscous type must be added in the mechanical sector of the model, i.e. that the balance equation for the angular momentum must be completed with an exchange term proportional to the angular velocity (Fig. 6b).

Also in this case, the proportionality constant (which depends mainly on the strength of the magnets and the distance between them and the copper plate – typically between 1 and 2 mm) is fixed through the comparison between the experimental data and the measured values, as shown in Fig. 6a: the mechanical part is now modelled in a satisfactory way.

This suggests asking students to consider the experiment also from the energy point of view: in fact, all aspects can be calculated without adding any new element, besides the constitutive relations that define the various energy terms. Fig. 6b shows the model extended to consider these aspects, while Fig. 6c shows the model predictions. The energy exchange associated with every process is calculated *separately* from its own constitutive law. It can be noted that the "mechanical" dissipation is less than 10% of the "magnetic" one, and also that the *total energy* – calculated as the sum of the rotational energy and the dissipated energy – turns out to be constant: this tells us that the model is coherent with the energy conservation principle.

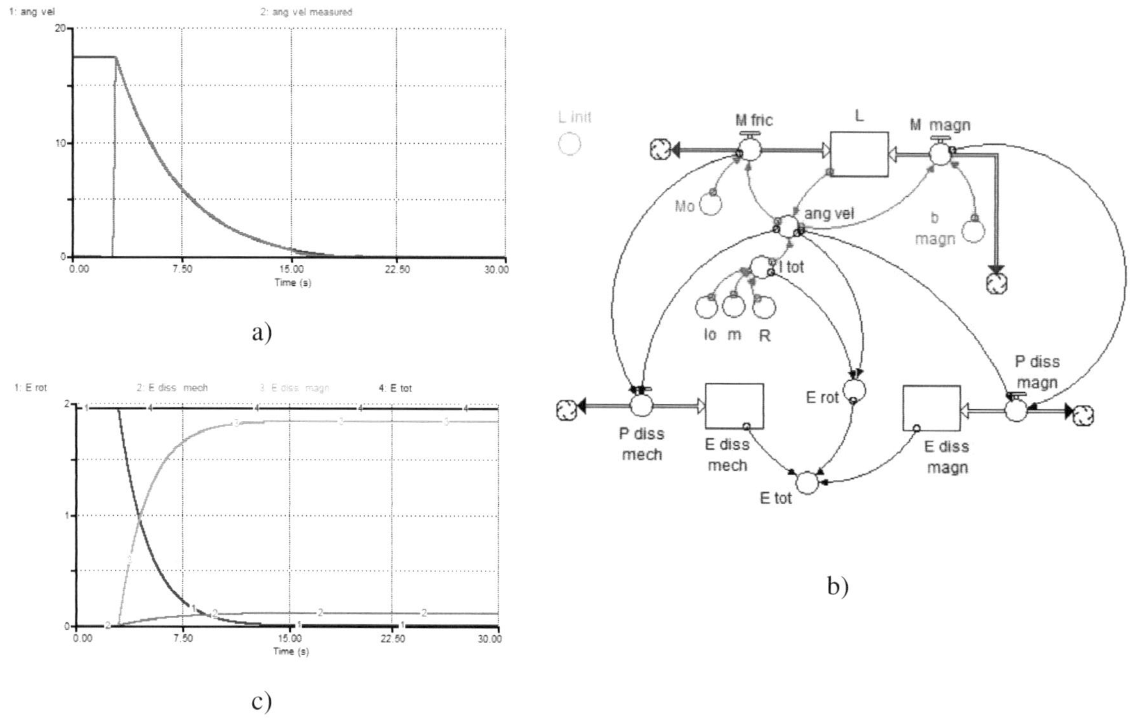

Fig. 6. A rotating disk braked by dry and magnetic friction: (a) angular velocity vs. time: simulated (1 – blue) and measured (2 – red) values; (b) model including energy aspects; (c) simulated values of the energy vs. time: rotation energy (1 – blue), energy dissipated by dry (2 – red) or magnetic friction (3 – green). The sum of the three contributes represents the total energy of the system (4 – black)

At this point the two parts of the model can be easily coupled, recognizing that *only* the magnetic interaction contributes to the temperature rise in the copper plate. Furthermore, we assume that the dissipative processes in the neodymium magnets can be neglected, due to their high magnetic stiffness: in order to test these ideas, in the model it is sufficient to interpret the power dissipated in the magnetic interaction as an additional exchange (inflow) in the balance equation of the (internal) energy of the copper plate (Fig. 7a). Fig. 7b shows the result for the change in temperature (comparison between measured data and model predictions).

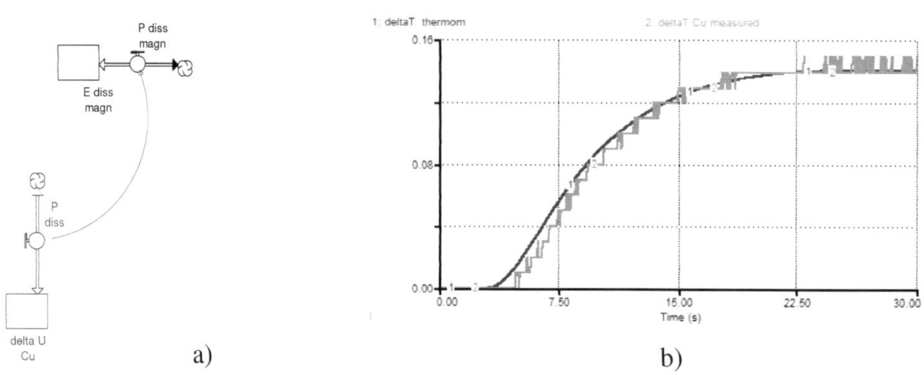

Fig. 7. A rotating disk braked by dry and magnetic friction: (a) the model: *only* the magnetic interaction contributes to the temperature rise in the copper plate; (b) temperature of the copper plate during the braking vs time: simulated (1 – blue) and measured (2 – orange) values

IOP Conf. Series: Journal of Physics: Conf. Series **1076** (2018) 012017 doi:10.1088/1742-6596/1076/1/012017

4. Conclusions

We presented and discussed an experiment with a rotating disc braked magnetically. As far as the mechanical aspects are concerned, our work confirms the principal results of previous studies of several aspects involved in rotational motion affected by friction (Amrani 2006, Eadkhong *et al.* 2012, Mungan 2013, Kladivova *et al.* 2016 and references therein), but it also allows us to extend the study to the thermal aspects, localizing the dissipative processes. Our model is, in fact, constructed step by step from the mechanical *and* thermal constitutive laws of interacting parts, fixing in separate experimental situations all the model parameters without the necessity to introduce any hypothesis concerning energy. This way, the resulting model can be seen as a real experimental test of the properties of energy itself, in particular *energy conservation*: the agreement of the model predictions with the whole set of measured data (relative to both the mechanical and thermal behaviour) shows, therefore, the robustness and the coherence of the whole model, not only as a mathematical structure, but above all as the result of the general approach used to describe energy exchanges.

References

Amrani D. (2006) Computerized rotational system to study the moment of inertia of different objects, *European Journal of Physics* **27**(5), 1063-1069.

Corridoni T., D'Anna M. & Fuchs H.U. (2014) Damped mechanical oscillator: Experiment and detailed energy analysis. *The Physics Teacher* **52**, 88-90.

D'Anna M. (2016) Addressing some common difficulties in teaching and learning energy in high school. *Girep Book of selected papers presented at the 2016 GIREP Seminar.* Focusing on Lab to improve Physics Teaching and Learning. Research based proposals (Sokolowska D., Michelini M., eds.), Springer.

D'Anna M. & Rosenberg J. (2014) Experiments. In Analogies: a key to understand physics. *Girep Book of selected papers presented in the GIREP-ICPE-MPTL International Conference* (Herrmann F. *et al.*). Teaching and Learning Physics Today: Challenges? Benefits? (Kaminski W., Michelini M., eds.), Reims, pp. 137-142. http://www.youtube.com/view_play_list?p=9B638811E3 6695C4.

Eadkhong T., Rajasadorn R., Jannual P. & Danworaphong S. (2012) Rotational dynamics with Tracker, *European Journal of Physics* **33**, 615-622.

Fuchs H.U. (2002) *Modeling of uniform dynamical systems.* Zürich: Orell-Füssli.

Kladivova M., Kovalakova M., Gibova Z., Fricova O., Hutnikova M. & Kecer J. (2016) Laboratory experiment for the study of fiction forces using rotating apparatus. *European Journal of Physics* **37**, 1-15.

Mungan C.E. (2012) Frictional torque on a rotating disc, *European Journal of Physics* **33**, 1119-1123.

IOP Conf. Series: Journal of Physics: Conf. Series **1076** (2018) 012018 doi:10.1088/1742-6596/1076/1/012018

Conceptual and exploratory labs for secondary teacher education in two different countries. The case of dc circuits

Jenaro Guisasola[1], Marisa Michelini[2], Alberto Stefanel[2], Kristina Zuza[1]

[1]Applied Physics Department, University of the Basque Country (UPV/EHU), Spain
[2]URDF University of Udine, Italy

Abstract. We analyse the effectiveness of our Formative Intervention Module (FIM), where our was to support a groups of 46 pre-service technology secondary teachers at the University in Udine and a group of 32 pre-service technology teachers at the University of the Basque Country. The topic of the formative intervention was DC circuits because it is a topic that is mentioned in the technology curriculum of any country. The research problem considers how we succeeded in promoting formative modules by promoting new learning materials and by organizing pre-service training. In particular we evaluate the pre-service teachers achievement both in understanding the concepts and in the acquisition of pedagogical content knowledge. A qualitative pre/post-test was designed to get data and analyse the impact of the formative intervention module. We find a generalized improvement in the macro-physical description of a simple DC (Direct Current) circuits. In addition, around two third of future teachers recognized learning difficulties for understanding concepts and the alternative conceptions of students.

1. Introduction

There is a big amount of research on students' difficulties and alternative conceptions in learning DC circuits in Secondary and University levels. Physics Education research has established well Secondary students difficulties regarding the learning of a macroscopic model of how dc circuit works. The literature has showed that a significant portion of Secondary teachers has also problems with a global scientific understanding of the DC circuits. Sometimes the teachers confuse the concepts of current and potential difference (cita), the influence of the topology of the system or, they have difficulties in applying the energy conservation. However, there are few studies that propose and evaluate formative courses for teachers for overcoming the mentioned problems. Moreover, we notice that the teaching and learning problems on DC circuit are concerned not only to the science curriculum but also to the technology curriculum, because the topic is in both. Often, technology is taught using the "black box principle": the students first finish the technological project and after, they learn the concepts and laws related to the project. In technology, practical applications may overcloud the physical concepts and laws that are behind the technological project or machine (Yager 1996).

A Formative Intervention Module (FIM) was designed taking in mind the demands of new standards in Science and Technology and, previous results of the educational research in DC circuits (Cohen *et al.* 1983, McDermott 1991, McDermott & Shaffer 1992, Hart 2008, Glauert 2009, Gunstone *et al.* 2009). The first main goal is related with conceptual expertise of the future teachers on conceptual knowledge on simple DC circuits highlighting the importance of the interpretative model. The second goal is related with the didactic perspective and the acquisition of pedagogical content knowledge. The FIM was implemented in two countries in the University of Udine (46 pre-service low secondary teachers) and in the University of the Basque Country (25 pre-service low secondary teachers). One of the tools for research our goals were to create teaching materials that would allow to study the basics of DC circuits, following the educational path proposed by us in previous studies (Testa & Michelini 2007, 2008; Guisasola *et al.* 2008). We included in the teaching materials conceptual and explorative labs activities. Through labs activities pre-service teachers, have the opportunity to understand what happens in an electrical DC (Direct Current) circuit and build the model of a circuit and, at the same time, acquiring scientific skill (Testa & Michelini 2008). It is underlined the importance of the educational labs where future teachers can directly be involved and immersed in an educational path to experience the phenomenological exploration of the phenomena (Michelini 2004, McDermott *et al.* 2000, Sokoloff *et al.* 2004, Eylon & Bagno 2006). It is essential the transition between the experiment results and the modeling of the phenomena (Hart 2008, Gilbert & Justi 2016, Windschitl *et al.* 2008). We feel that technological curriculums should include the knowledge in science and technology, linking the learning of knowledge to the acquisitions of skills (Koballa *et al.* 1997). Working with an electric circuit project, for example, may improve general understanding of model of current and working skills that allow bringing the experimental data and explicative models.

The FIM was planned keeping in mind the results from studies made by Shulman (1986) and Roth (2007), between others that contributed to define the Pedagogical Content Knowledge (PCK) as the knowledge that teachers develop to help others to learn and that they build according to the specific topics of their area of knowledge. This knowledge includes not only the knowledge of the discipline, but the analysis and reflection about learning objectives and active experiences of educational paths (Gess-Newsome 1999, Magnusson *et al.* 1999, Etkina 2010, Guisasola *et al.* 2013, Michelini *et al.* 2013). One of the tools for working pre-service teachers PCK was to create a package of task that would allow reasoning ways of learning the foundations of the DC circuits according with the curriculum teaching objectives and their and/or students' learning difficulties.

The FIM was designed looking to the competencies that future teachers have to develop during their formation and other aspects as attitudes or STSE aspects were included in the design. The proposed activities are grounded on active learning strategies giving to the Pre-Service Teachers (PT) opportunities for building and discussing of DC circuit models to work in the objectives presented in the table 1.

Table 1. Relation of the content related objectives of the FIM

Objective	
1	Explicative model of simple resistive DC circuit
2	Distinguishing series and parallel conditions with simple DC circuits and functioning according different perspectives (e.g. topological perspective, Current-resistance-Tension one)
3	Overcome the main conceptual knots as those on the functioning of the circuit and topological one (e.g. equivalent circuits)
4	Deeper explicative model: attribute role the battery; meaning of battery as a tension generator

In addition to developing the study materials, the pre-service teachers (PT) were given training in the teaching of the DC circuits. This training was organized during a one course for science and engineers who wanted to become Secondary science and technology teachers. The characteristics of the course are described in the next section.

The research problems of the presents study are linked to the principal goals of the FIM. There are two main research questions:

- To what extent have we succeeded in the production of new materials and the organization of training in relation to the pre-service teachers' understanding of the scientific model of electrical circuits?
- To what extent have teachers acquired the PCK through the use of course materials?

2. The study

To answer the proposed research questions we implemented the FIM in two universities (University of Udine, UD) and (University of the Basque Country in Donostia, UPV/EHU-DO) in two different European countries (Italy and Spain). Although the two courses are not exactly the same there were many similarities concerning, for instance, educational approach, based on problem-solving and the experiments, reflections on conceptual knots and, development of conceptual model. The most important is that the learning goals concerning conceptual knowledge and PCK of both courses are very similar and that implies that both PT courses are comparable. The course done following the FIM in the University of Udine (4 ECTS) was implemented twice; at all 46 PT were enrolled aged between 30 and 60. In the University of the Basque Country the course (3 ECTS) was implemented once and there were enrolled 32 PT aged between 23 and 50. Table 2 resumes the degrees of the PTs in the two contexts.

Table 2. Degrees of the PTs involved in the two courses in Udine and Donostia

Degree	UD1 (N1 = 27)	UD2 (N2 = 19)	DO = UPV/EHU (N3 = 32)
Engineering	3	1	20
Architecture	21	17	6
Computer Science	1	1	6
Agronomy	1		
Chemistry	1		

The approach was the same in both courses adopting similar teaching. The course in Udine was organized as an educational lab including tutorials, experiments, interactive lectures demonstrations, design activities, questionnaires (Testa & Michelini 2007). The course in the UPV/EHU adopted an Interactive Teaching methodology through Project/Problem based learning, without "lecture", with lab experiments imbibed into the classroom activities. In appendix I some examples of analogous task that were implemented and adapted in the two context are given. Table 3 summarizes both teaching approaches.

Table 3. Schema of the Teaching Educational Approach, indicating, per each experiment, subject and typology of activity carried out (PS-Q: problem solving based on qualitative analysis; SQ: Semi-quantitative experiment; Exp: quantitative measurement-experiment; Exe: exercise). "U" means University of Udine and "D" means University of the Basque Country

Experiment	Subject	PS-Q	SQ	Exp	Exe
Circuit with bulb, battery, wire	Closed circuit → e. Current	U; D			
Circuit with bulb, battery, wire, different objects	Insulator; conductors, semiconductors	U	U		
Circuit with battery, wire and bulb in different positions → Ammeter, Voltmeter	The current is the same in all the section of a series circuit powered in continuous Tension is distributed	U; D	U; D	U; D	
Circuits and a series of batteries	The tension is an additive quantity	U	U		
Circuit with Parallel/series bulbs	The brightness → the current → Tension distribution, nodes law	U; D	U; D	U; D	U; D

	Systemic nature of circuits Equivalent load	U; D	D	U; D
Circuit with Parallel/series LED-		D	D	D
Logic Circuit	Equivalent functioning and topological differences	U; D		
Characteristic current-tension for a metallic wire	Ohm (DV = RI), (R = ρL/S) laws		U; D	

To analyze the effectiveness of the materials of the FIM and answer the research questions, a questionnaire composed by 4 questions was designed to administrate as pre-test and post-test instrument in both universities. In table 4, the relation between the objectives of the FIM and the research questions and the questions is shown.

Table 4. Objectives involved in each questions of the test and RQ focussed

Objective Research Question	**QuestionQ1**	**Question Q2**	**Question Q3**	**Question Q4**
Objective O.1.	x			
Objective O.2.		x		
Objective O.3.		x		x
Objectiv O.4.			x	
Research Question1.		x	x	x
Research Question2.	x			

The questionnaire is physics education research informed and is based on the learning difficulties and reasoning detected in previous research works (Cohen *et al.* 1983, McDermott 1991, McDermott & Shaffer 1992, Hart 2008, Glauert 2009, Gunstone *et al.* 2009). In appendix II there are the four questions. For the internal validity of the questionnaire three independent teacher who were experts in the teaching the topic, were asked to respond the questions and to analyze the objectives of the questions. They made suggestions that were taken into account when writing its final version. All faculty members confirmed that the contents of the questionnaire and the objectives were appropriate. Additionally, a pilot study was conducted with small samples of students. This confirmed that students generally had no problem understanding the meaning of questions.

3. Data analysis and results

The analysis of each question was made twice from two different points of view. The first analysis was done regarding to the final answer and they were classified taking into account the correctness of the answer. In the second analysis the interpretative approach was studied. PT' answers were analyzed independently by two researchers; Cohen's kappa reliability coefficient averaged 0.86 for the questions, indicating very good concordance in the judges' criteria for setting the categories described. The intra-rater reliability kappa coefficient was also calculated for the main researcher three weeks later, obtaining a value of 0.87, on average, for all the questions, which is satisfactory for a level of confidence of 95%. The results of these semi-quantitative analysis are shown in frequencies comparing pre-test and post-test situations and different cohorts for Udine and The Basque Country.

We will present some results from three questions. In the first question Q1 of the questionnaire a simple situation is proposed (See Appendix II: Q1). A circuit is formed by a 4.5 V battery, connection wires, a bulb, a switch. When you close the circuit, the light bulb turns on. The PTs were requested to discuss the model that can explain this fact. In the second part of the question Q1, they have to discuss some typical student responses and illustrating the reasoning that underlies each model. PTs answers were classified according to the qualitatively different criteria to analyze the students reasoning (see table 5). We defined four categories of explanation for the Q1.2:

A. *Scientific:* The PT explains the scientific concept involved, don't considering the sentences of students
B. *Correction:* The PT corrects the students answer, evidencing often what is wrong
C. *Correct/incorrect:* The PT indicates when the answer is correct or incorrect (without explain why)
D. *Student reasoning:* The PT indicates the students reasoning (in his opinion)

The results of question Q1.2 obtained in pre-test and post-test in both samples are shown in table 6. In this case the explanation given by students for each model proposed were analyzed and categorized in one of the explanation described before. We found that in both universities have grown two categories (correction on the answer and student reasoning) and "correct/incorrect" and "No answer/no explanation" (NA/NE) categories have decreased.

Table 5. Percentage concerning PTs analysis of students responses proposed in Q1

Q1.2	A. Scientific explanation		B. Correction		C. Incorrect		D. Student reasoning		E. NA/NE	
	Pre	Post	Pre	Post	Pre	Post	Pre	Post	Pre	Post
UPV/EHU	7.0	3.0	40.0	77.0	35.0	0.0	15.0	20.0	3.0	0.0
U. Udine	0.0	6.5	28.0	32.5	15.5	3.0	11.0	41.0	45.5	15.5

In table 5, the percentages of pre-test students' responses showed problems in learning knots concepts for the UD cohort (45.5% of the total), due probably to the poor scientific baggage of the PT's. The answering PT's usually addressed in terms of accuracy / inaccuracy of the responses of the students (15,5-35% in category C) or indicating in what respects the response was scientifically wrong (28-40% in category B).

In the post-test, 41% of UD cohorts indicate the reasoning underlying the responses of the students and 20% of UPV/EHU cohort (see category D). However, in University of Udine there is a higher percentage of NA/NE than in UPV/EHU, after teaching the FIM. These different percentages in the groups of UD and UPV/EHU reflects, firstly, the different backgrounds of PTs (basically architecture for UD cohorts; engineering for UPV/EHU cohort), and, secondly, the more number of tasks devoted in UD to the analysis of the conceptual problem of pupils. 77% of UPV/EHU cohort and 32.5% of UD cohorts explain in what sense the students' response was not correct scientifically (see category B). 7.5% provides an analysis on the scientific level of the situation and 3.5% simply indicate whether the students' answers are correct or incorrect (see category C), which is an explanation more superficial of the explanations from category B.

In the two contexts, significant changes are evident comparing pre/post results. In addition to the change in the percentages, there are changes in relation to the quality of the explanations. The vast majority of the post-test answers are in the categories A, B and C. On the one hand, the categories A and C (18.0% UPV/EHU and 17.5% UD) are principally related to the scientific understanding of the DC circuits and the PTs' explanations critics the different options of the question Q1.2 from this point of view. On the other hand, a significant percentage of answers analyze each sentence of the question Q1.2 from the point of view of PCK taking in mind students alternative models (77% UPV/EHU and 32.5% UD). In total, the vast majority of the answers in the post-test show PTs' progression in learning DC circuits explicative model. Some standard examples of the progression in both understanding and acquisition of PCK are showed following:

There is current flow only from the positive pole to the bulb. There is no current flow that comes out from the base of the bulb since the current is used to turn on the bulb. For circulating current is necessary to have a closed circuit (understanding of the model of the circuits, Q1.1)

The same student explaining the options of Q1.2 writes:

Model A and B suppose no circulation of the current for all circuit. According to these interpretations is implicit that the bulb uses all the current or that the bulb sucks current from each pole. Model C supposes that the electric current go form positive pole to negative but is not correct that the current decrease after pass through the bulb. Although the bulb consumes energy, but the current is the same in all circuit. Model D is the correct answer because users do not consume current

In question Q2, three bulbs connected in parallel in an unusual way are presented (see Appendix II). In the first part of this question, PTs are request to compare the brightness/current of each bulb. In the third part, the bulb 3 is short cut and PTs have to answer if something changes due to that short cut.

Table 6. PTs percentage of answers on the question posed in Q2.1. The expected answer was i1 = i2 = i3

Q2.1	L1 > L2 > L3		L3 > L2 > L1		L3 ≠ L2 = L1		L1 = L2 = L3 (i1 = i2 = i3)		NA/NE	
	Pre	Post	Pre	Post	Pre	Post	Pre	Post	Pre	Post
UPV/EHU	0.0	0.0	0.0	0.0	81.0	51.5	15.0	48.5	4.0	0.0
U. Udine	4.5	0.0	6.5	0.0	26.0	41.0	19.5	56.5	43.5	2.5

Table 6 reports the percentage of PTs answers to the question Q2.1. In the pre-test the answering prevalently do not recognize the equality of brightness/current. Moreover, in Udine almost an a half do not answer. In the post about a half of the answers in both UPV/EHU and UD, gave the expected answer (48.5% UPV/EHU and 56.5% UD). However, a significant part of PT evidences difficulties in recognizing the topology of the circuit and in particular the equivalent role of L3, with respect to the role of L1 and L2 (51.5% UPV/EHU and 41.0% UD).

Standard examples of the answers in the pre-test for category L3 ≠ L2 = L1 is the following:

The bulbs l1, l2 are connected in parallel, the L3 bulbs with the L1 + L2 group also in parallel. The brightness will be different: L1 = L2 but less than L3.

L1 and L2 equal to each other. L3 different from L1 and L2 (+ light); The I is divided into 2 point A (and becomes i1 and i2); i1 is still divided into i3 and i4.

This type of reasoning remains the same for the answers in the post-test.
A standard example of the correct answer in the post-test is:

Equal brightness because they are in parallel. That is, we have the same voltage in L1, L2, L3. while the current is distributed.

The vast majority of the answers in the pre-test use typical local reasoning, based on incorrect assumptions, which led to wrong answers (as see in table 6). In contrast, in post-test there is a significant improvement in understanding the circuit as a system.

Question 4 is similar to the question Q2 but in an academic context familiar to PT. Regarding to the final answer and if we compare with question Q2.1, the success of the students is much better. Even in the pre-test the answer are quite good (52% in Udine and 88% in UPV/EHU). In the post-test almost

100% gave the correct answer. In table 7, the comparison of the answers of questions Q4 and Q2.1 is presented regarding to the line of reasoning used in the answers, that includes conceptual comprehension of the aspects listed before. Here we stress on the following points: the Explanation A is given in the Q2.1 post-test by 48-30.5% of PTs and only by 15.5% in the Q4 post-test; the equal tension for each bulbs (absent in the pre-test) was given by 52% in UD and 16% in UPV/EHU; other naïve answers pass from 41/51.5% to 10.5/22%. It is evident the different explanation perspectives used, according to the different context considered.

Table 7. Comparison of PTs percentage of answers to question Q2.1 and Q4

Q2.1 and Q4	Udine Q2.1		UPV/EHU Q2.1		Udine Q4		UPV/EHU Q4	
	Pre	Post	Pre	Post	Pre	Post	Pre	Post
A. Bulbs/resistance in parallel	9.0	48.0	11.0	30.5	19.5	15.5	18.0	15.5
B. Nodes law ($i_{in} = i_{Out}$)	6.5	9.0	41.0	18.0	6.5	20.0	4.0	19.0
C. Equal current	0.0	0.0	0.0	0.0	2.0	2.0	37.0	28.0
D. Different length of the arms	6.5	2.0	0.0	0.0	11.0	0.0	7.5	0.0
F. Equal ΔV	0.0	0.0	0.0	0.0	9.0	52.0	22.0	15.5
G. Others	78.0	41.0	48.0	51.5	52.0	10.5	11.5	22.0

4. Conclusion and implications

We have noted the demands of new standards in Science and Technology. We have considered a common topic (electric circuits) in both the curriculum in science and technology to develop a formative intervention module for Pre-service Teachers (PT) who will teach technology and/or science in secondary education (12-16 years old students). It came as a surprise that the knowledge of cohort of engineers and architects from both countries, reacting spontaneously to pre-test, gave very poor arguments and show poor understanding on DC circuits.

A FIM on DC circuits based on exploratory and conceptual activities was designed for the preparation of secondary pre-service teachers of technology. Furthermore, this module was implemented in two universities of Europe: University of Udine (Italy) and UPV/EHU (Basque Country, Spain). Two courses were implemented adopting a similar operative approach, adapted in two different national contexts. The aim of the FIM is to improve the competencies of the prospective teachers in the interpretation of the topology and functioning of DC circuits, using physics concepts to construct interpretive model, in a way that allow them to provide to pupils the understanding on the topic of DC circuits.

As instrument for data collection we used a pre/post-test concerning the learning conceptual and didactic objectives of the FIM and based on open questions. The analysis of the PTs responses, give us the opportunity to answer to the research questions.

Concerning how the organization of training with the new materials improve the understanding pre-service teachers the understanding of the main conceptual aspects of DC circuits (RQ1), we find a generalized enhancement in the macro-physical description of a simple DC circuit using basic concept as current, potential difference, resistance, the equivalence of the circuit and batteries. More precisely, from pre-test to post test the PTs have demonstrated a higher level of skill acquisition. One of the best examples of that is the evolution of the percentage of correct answer and PTs that do not answer the questions.

Concerning how the teaching educational approach contribute in gaining PCK competencies on DC electrical circuits (RQ2), prospective teachers changes the ways of analyzing the conceptual problems related to the conceptions at the base of students' reasoning (almost 2/3 in Q1). PTs overcome the initial tendency of teachers to evaluate the student answer only as correct/incorrect. It emerges the competence on critical analysis on the scientific point of view. The majority of PTs passed from the manipulation of circuit to conceptual construction of concepts using an operative/experimental approach.

The outcome of the FIM is good when compared with the results of the traditional teaching in other studies. This study reveals information about the attainment of project goals, which are to help PT to acquire, on the one hand, the concepts and skills in the basis of explanatory model of DC circuits and, on the other hand, the necessary PCK for a good teaching of the topic. However, the study does not give information about the pre-service teachers teaching strategies when they teach the topic in Secondary school. Therefore, further research will be necessary aiming to determine the pre-service teachers educational practice when they teach the topic.

References

Cohen R., Eylon B. & Ganiel U. (1983) Potential difference and current in simple electric circuits: A study of students' concepts. *American Journal of Physics* **51**, 407-412.

Etkina E. (2010) Pedagogical content knowledge and preparation of high school physics teachers, *Physical Review Special Topics Physics Education Research* **6**, 020110.

Eylon B.-S., Bagno E. (2006) Research-design model for professional development of teachers. *Physical Review Special Topics Physics Education Research* **2**, 020106.

Gess-Newsome J. (1999) PCK: an introduction and orientation. In *Examining pedagogical content knowledge* (Gess-Newsome J. &. Lederman N.G, Hrsg.), Dordrecht: Kluwer, pp. 3-17.

Gilbert J.K. & Justi R. (2016) *Models and Modeling in Science Education* **9**, Dordrecht: Springer.

Glauert E.B. (2009) How young children understand electric circuits: Prediction, explanation and exploration. *International Journal of Science Education* **31**, 1025-1047.

Guisasola J., Barragués J.I. & Garmendia M. (2013) El Máster de Formación Inicial del Profesorado de Secundaria y el conocimiento práctico profesional del futuro profesorado de Ciencias Experimentales, *Matemáticas y Tecnología, Revista Eureka sobre Enseñanza y Divulgación de las Ciencias* **10**, 568-581.

Guisasola J., Michelini M., Mossenta A., Viola R (2008) Teaching electromagnetism: issues and changes. In *GIREP-EPEC Conference Frontiers of Physics Education 2007. Selected Contributions* (Jurdana-Šepić E. *et al.*, Eds.), Zlatni rez: Rijeka, pp. 58-76.

Gunstone R., Mulhall P. & McKittrick B. (2009) Physics teachers' perceptions of the difficulty of teaching electricity. *Research in Science Education* **39**, 515-538.

Hart C. (2008) Models in physics, models for physics learning, and why the distinction may matter in the case of electric circuits. *Research in Science Education* **38**, 529-544.

Koballa T., Kemp A. & Evans R. (1997) The spectrum of scientific literacy. *Science Teachers* **64**, 27-31.

Magnusson S., Krajcik J. & Borko H. (1999) Nature, Sources, and Development of PCK for Science teachers. In *Examining PCK* (Gess-Newsome J. & Lederman N.G., eds.), Dordrecht: Kluwer, 95-132.

McDermott L.C. (1991) What we teach and what is learned: Closing the gap [Millikan Lecture 1990]. *American Journal of Physics* **59**, 301-315.

McDermott L.C. & Shaffer P.S. (1992) Research as a guide for curriculum development: An example from introductory electricity. *American Journal of Physics* **60**, 994-1003.

McDermott L.C., Shaffer P.S., Constantinou C.P. (2000) Preparing teachers to teach physics and physical science by inquiry, *Physics Education* **35**(6), 411-416.

Michelini M. ed. (2004) *Quality Development in the Teacher Education and Training,* selected papers in Girep book, Udine: Forum.

Michelini M., Santi L. & Stefanel A. (2013) La formación docente: un reto para la investigación, *Revista Eureka sobre Enseñanza y Divulgación de las Ciencias* **10**, 846-870.

Shulman L.S. (1986) Those who understand: knowledge growth in teaching. *Educational Researcher* **15**(Z), 4-14.

IOP Conf. Series: Journal of Physics: Conf. Series **1076** (2018) 012018 doi:10.1088/1742-6596/1076/1/012018

Roth K.J. (2007) *Science Teachers as Researchers*. In Handbook of Research on Science Education (Abell S.K. & Lederman N.G., eds.), N.Y.: Routledge, pp. 1205-1259.

Sokoloff D.R., Lawson P.W., Thornton R.K. (2004) *Real Time Physics*, New York: Wiley.

Testa I., Michelini M. (2008) Supporting global reasoning in electric circuits: a functional approach to address common misconceptions about electric circuits. In *GIREP-EPEC Conference Frontiers of Physics Education 2007. Selected Contributions*.(Jurdana-Šepić E. *et al.*, eds.), Zlatni rez: Rijeka, pp. 77-80.

Testa I., Michelini M. (2007) Prospective primary teachers 'functional models of electric and logic circuits. In *Modelling in Physics and Physics Education* (Van den Berg E., Ellermeijer T., Slooten O., eds.), Amsterdam: University of Amsterdam, 391-404.

Windschitl M., Thompson J. & Braaten M. (2008) Beyond the scientific method: Model-based inquiry as a new paradigm of preference for school science investigations, *Science Education* **92**, 94.

Yager R.E. (1996) Science/Technology/Society as Reform in Science Education. Albany: State University of New York.

IOP Conf. Series: Journal of Physics: Conf. Series **1076** (2018) 012018 doi:10.1088/1742-6596/1076/1/012018

APPENDIX I

Examples of activities used in the FIM implemented the University of Udine and in the University of the Basque Country. A) Circuits with a battery and a bulb used in the Udine and Donostia courses. B) Circuits connecting a battery to parallel/series bulbs and C) examples of operative introduction of the role of battery in a circuit (tension generator), comparison of the bulbs brightness.

University of Udine UPV/EHU

A

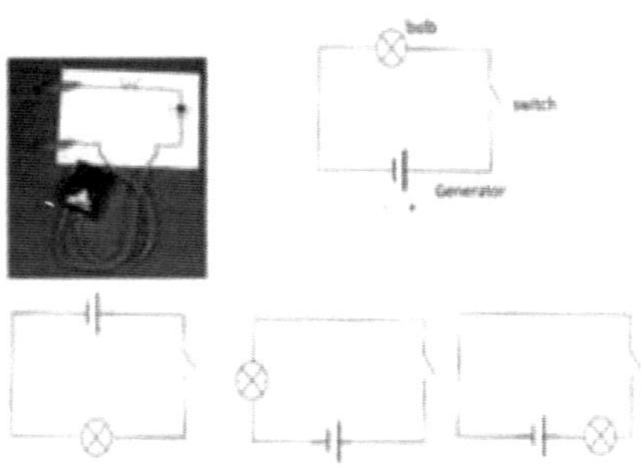

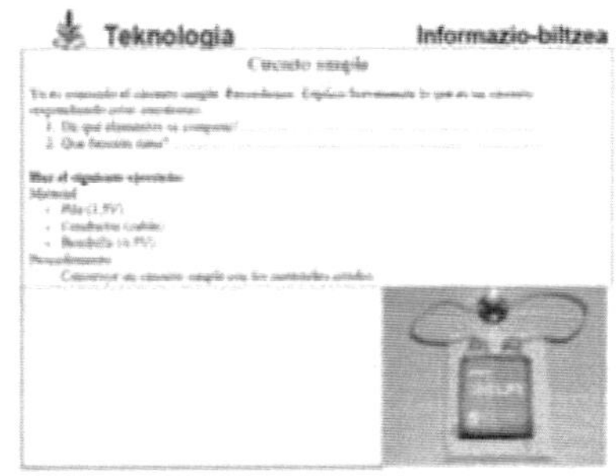

B

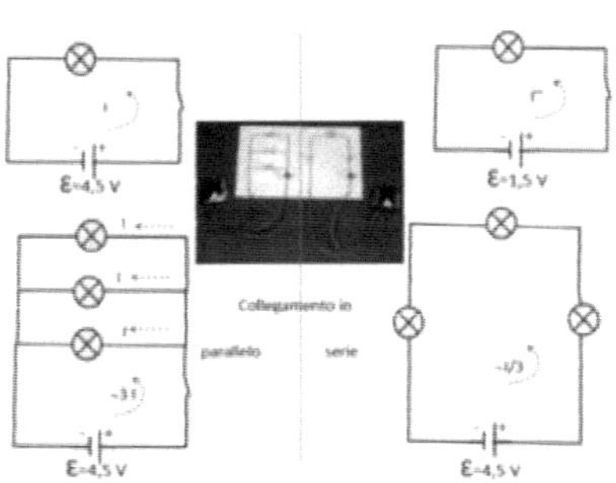

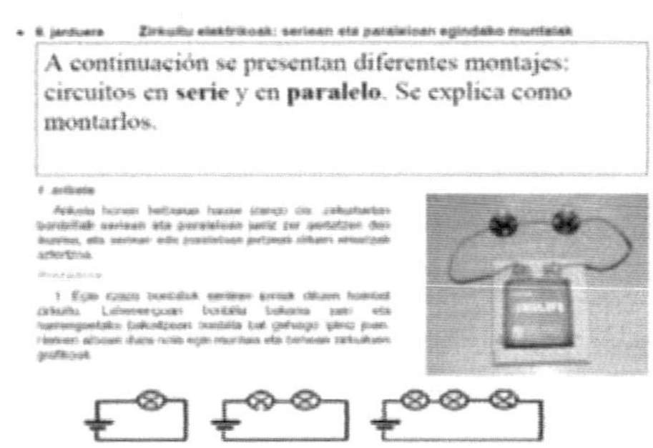

C

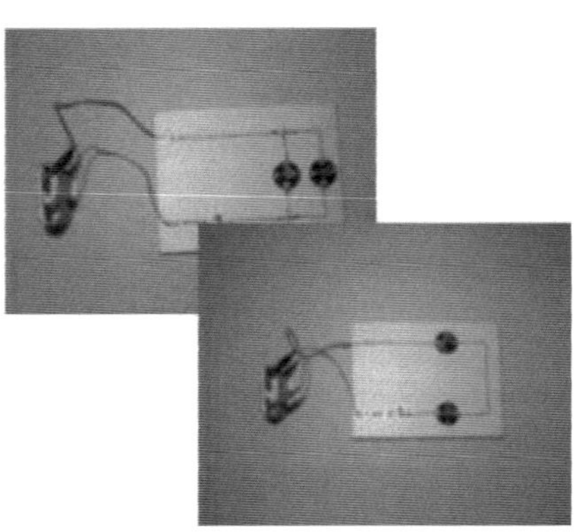

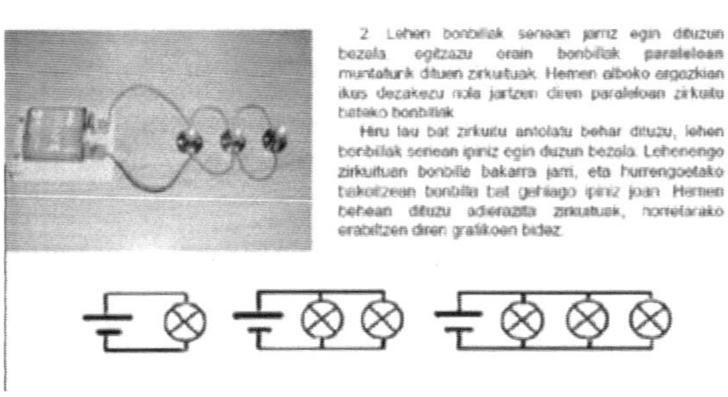

IOP Conf. Series: Journal of Physics: Conf. Series **1076** (2018) 012018 doi:10.1088/1742-6596/1076/1/012018

APPENDIX II

Q1. A circuit is formed by a 4.5 V battery, connection wires, a bulb, a switch. When you close the circuit, the light bulb turns on.

1.1 Which is the model that can explain that?

....

....

....

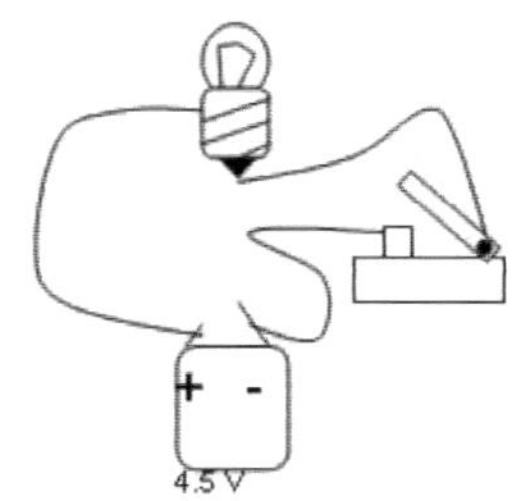

1.2 Please, discuss some typical student responses, illustrating in particular the reasoning that underlies each model:

Model A: There is current flow only from the positive pole to the bulb. Any current flows from the base of the bulb since the current is used to turn on the light bulb

Model B: The current flows from each of the two poles of the battery to the bulb (it lights up for the clash of the two currents in the opposite direction)

Model C: The current circulates always in the same direction around the circuit, but his intensity is lower after the bulb

Model D: the current circulates in the same direction and with the same current throughout the circuit.

Q2.

Q2.1 What will be the brightness of each of the bulbs (I1, I2, I3, respectively)? Explain

Q2.2 Changes the brightness of the bulbs if unscrewing L3? Explain

Q2.3 Changes the brightness of the bulbs if you short L3? Explain

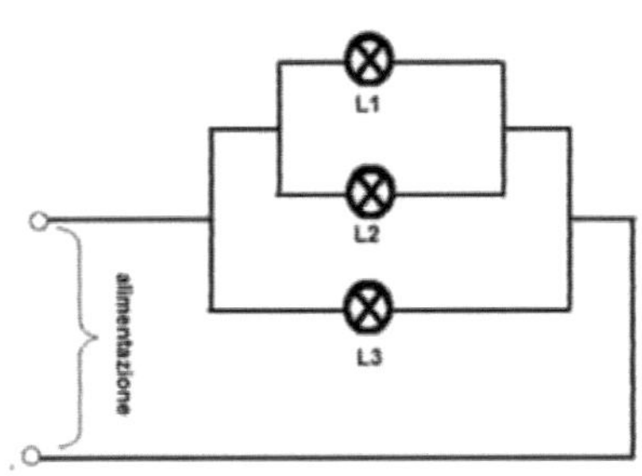

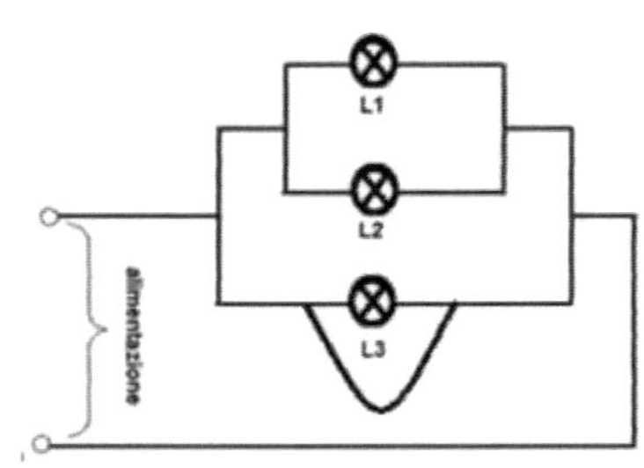

Q4. Consider the circuit formed by three identical bulbs connected in parallel, a switch and a battery.

When the circuit is closed, the brightness of each bulb will remain the same?

Explain your answer.

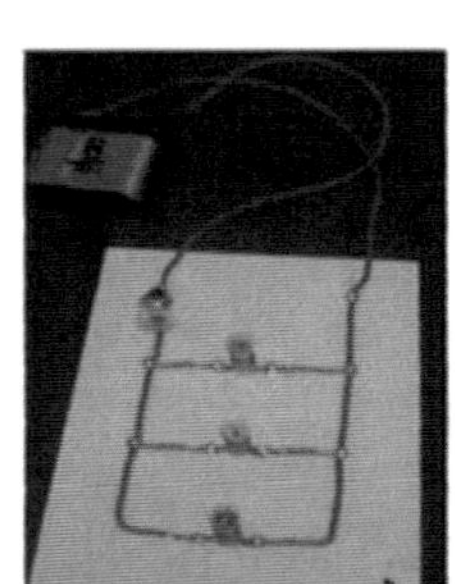

IOP Conf. Series: Journal of Physics: Conf. Series **1076** (2018) 012019 doi:10.1088/1742-6596/1076/1/012019

A quantitative analysis of university student reasoning lines in the field of thermally activated phenomena

Onofrio Rosario Battaglia[1], Benedetto Di Paola[2], Claudio Fazio[1], Nicola Pizzolato[1], Dominique Persano Adorno[1]

[1]UOP_PERG (University of Palermo Physics Education Research Group), Dipartimento di Fisica e Chimica, Università di Palermo, Italia
[2]GRIM (Mathematics Education Research Group), Dipartimento di Matematica e Informatica, Università di Palermo, Italia

Abstract. In this contribution we present a research aimed at studying the effectiveness of two workshops in improving reasoning skills in undergraduate students. Both the workshops are based on the Feynman Unifying Approach. A questionnaire containing six open-ended questions on the temperature dependence of evaporation of a liquid and of a chemical reaction was administered to the students of both groups before instruction. A second one, similar to the first but focused on a physical content different from both the pre-instruction test one and the content dealt with during the workshops, was administered after instruction. The responses to the pre- and post-instruction questionnaires are analyzed by using Not-Hierarchical Cluster Analysis methods and students' lines of reasoning about the proposed phenomena/situations are inferred in both the experimental and the control group. The implications on the efficacy of the two workshops in improving student explicative skills are discussed.

1. Introduction

Students' descriptive and explanatory skill development and use is a relevant aim of university programs, for a meaningful understanding of science, as well as for the development of professional competencies. Research has shown the relevance of characterizing the mental models (Greca & Moreira 2000, Johnoson-Laird 2006) students use when asked to create or use explanations. It was shown (Clough & Driver 1986, Bao & Redish 2006) that students are often inconsistent in their use of mental models in situations that an expert would consider equivalent. As it is well known, in the construction process of explicative models inductive reasoning is involved, but an important role is also played by analogical reasoning (Duit & Glynn 1996). This involves the ability to see similarities and differences between a "source" (something perceived as similar to what we are going to analyze) and the "target" (the real phenomena that we are studying), and to generalize ideas and concepts already developed in a given context to different ones. This point is particularly relevant for undergraduate student science education, which aims to develop explicative and generalization skills in students, also by supplying them with unifying frameworks for the description and interpretation of natural phenomena only apparently different.

It is widely accepted that the need to identify differences and similarities between descriptive and explicative procedures, as well as the way these are used to generalize the results and account for real-world phenomena, can be met by introducing the students to the practice of scientific inquiry (AAAS 1993, Sadeh & Zion 2009, NRC 2012, Etkina 2014, Pizzolato *et al.* 2014). It would be interesting to study how the exposition of undergraduate students to an Inquiry-Based workshop focused on the well

IOP Conf. Series: Journal of Physics: Conf. Series **1076** (2018) 012019 doi:10.1088/1742-6596/1076/1/012019

know Feynman Unifying Approach (FUA)[1] to thermally activated phenomena (Feynman *et al.* 1963) can be effective in modifying the student lines of reasoning, redirecting them to explicative-like ones. Moreover, a comparison among these modified lines of reasoning and the ones that the students could develop by means of a different, more *traditional* approach to science learning, still based on FUA, but not explicitly focused on inquiry, would also be interesting.

In this paper, we discuss some of the results of a study involving a sample of undergraduate students in the second semester of their freshman year of the Undergraduate Program in Chemical Engineering. A questionnaire containing six open-ended questions on thermally activated phenomena was administered to the students of both groups before instruction. A second one, conceptually similar but focused on different physical content was administered after instruction. A quantitative analysis of the questionnaire responses was performed by using the k-means method (MacQueen 1967). It is aimed at allowing the researchers to group the students in intellectually similar subgroups (clusters) and at easily evidencing common patterns in the student responses to the questions. This procedure can help the researchers to infer the student lines of reasoning related to the creation and use of explanations in an unsupervised method (see, for example Sathya & Abraham 2013, Battaglia *et al.* 2017).

In the following sections, we present the research question addressed in this paper, a brief description of the analysis methods used and our results. Final comments about the implications of our results for the physics education of undergraduate students and suggestions for further developments are provided at the end of the paper.

2. Instruction and data collection

Based on the ideas discussed above, we chose to focus our workshops on the physics underlying the complex world of thermally activated phenomena, because it offers a good opportunity to understand and use unifying frameworks for the description and explanation of natural phenomena concerning apparently different fields of science. In particular, we focused on physics, chemistry and biology systems that can exist in two different states characterized by an energy difference ΔE (Boltzmann 1909a, 1909b, reissued 1969) where the state transition is thermally activated by overcoming the potential barrier ΔE. They are described by a unifying expression containing the Boltzmann factor, $e^{-\frac{\Delta E}{kT}}$, where T is the system temperature and k is the Boltzmann constant.

Research Question

Our study is centred on the comparison of two learning environments. In them, undergraduate students are involved in the construction and use of explanations of thermally activated phenomena in a context oriented to the development of a unifying approach to natural phenomena. Taking into account these considerations, we formulated the following research question for this study: *How are the lines of reasoning applied by undergraduate students when asked to make sense of situations related to thermally activated phenomena modified by two different learning environments both focused on a unifying approach to natural phenomena?*

Context, sample and methodology

Our research sample consists of 72 freshmen attending the Undergraduate Program in Chemical Engineering during the Academic Year 2014/2015 at the University of Palermo (UniPA), Italy. During the 1st semester of their Degree Program the students had attended general mathematics, physics and inorganic chemistry courses, and passed the exams. When selected to participate in our study, they were attending a 2nd semester Physics course dealing with the fundamentals of electromagnetism. The sample of students was randomly divided in an experimental group and in a control one (Myers & Hansen 2012), each made of 36 students. The 36 students of the experimental group attended

[1] According to this approach, phenomena apparently different can be described and explained by using a same conceptual framework, i.e. the idea of two-level system and the mathematical description involved by Boltzmann Factor.

a 20-hour, strictly inquiry-based workshop posing their questions, designing and carrying out their own investigations, gathering information, collecting and analyzing data, providing explanations and sharing the results. The other 36 students (control group) attended a more traditional workshop, still based on laboratory and modelling activities but they have never been requested to follow the typical steps of an inquiry-based approach.

The reasoning deployed by the students when asked to explain phenomena, and relate them to the physics and chemistry they had already studied in previous courses, was studied before instruction by using a previously validated (Fazio *et al.* 2013), specially designed questionnaire. In the questionnaire, students were asked: 1) to discuss a real life situation (the evaporation of a water puddle at different environmental temperatures); 2) to describe the physical quantities contained in Arrhenius' Law; 3) to clarify the role of a catalyst in a chemical reaction; 4) to give a microscopic interpretation of the Arrhenius' Law; 5) to show generalization skills by finding other natural phenomena that exhibit temperature dependencies similar to the one highlighted by the chemical reaction speed, and 6) evidencing the similarities among these phenomena, particularly with respect to common physical quantities characterizing all the described systems.

The students in the experimental group then took a 20-hour workshop based on a Bounded/Open inquiry-based approach (Wenning 2005). The workshop dealt with a physical content (electricity) different from the one addressed by the questionnaire, but strictly related to the framework of thermally activated phenomena. The students in the control group, instead, took a course of equal duration and with the same instructors of the previous one. During this course the same physical content was dealt with, but this time the pedagogical approach was a more traditional one, still based on laboratory activities but not explicitly focused on Inquiry.

At the end of the workshops, a new questionnaire, validated by following a procedure similar to the one used for the pre-instruction questionnaire, and again focused on the study of student lines of reasoning about the use of descriptions/explanations in science, was administered to the students of both groups. This questionnaire was conceptually similar to the pre-instruction one, but was focused on physical/chemical contents (fluidity) not explicitly discussed before and/or in the workshop. In this questionnaire, students were asked: 1) to discuss a real life situation (the flow of oil in pipes at different environmental temperatures); 2) to describe the physical quantities contained in the viscosity expression $\eta(T)$ according to Eyring's absolute rate theory; 3) to clarify the role of additives in oil industry; 4) to give a microscopic interpretation of the $\eta(T)$ Law; 5) to show generalization skills by finding other natural phenomena that exhibit temperature dependencies similar to the one highlighted by the $\eta(T)$ law and 6) evidencing the similarities among these phenomena, particularly with respect to common physical quantities characterizing all the described systems. All students in both groups completed the post-instruction questionnaire.

Specific content and Workshop description

The two workshops dealt with the study of electric current in materials (conductors and semiconductors) and in vacuum systems (thermionic tubes). In particular, situations where the Boltzmann Factor (BF) can be used to describe electric conduction were analyzed. The experimental group workshop was divided into five phases, specifically following the 5E- model approach. These phases are briefly described in table 1. Also the control group activity was divided into five phases. Some detail can be found in table 2.

Table 1. *Inquiry Based workshop*

Phase	Description of the activities	Hours
Engagement	Presentation of the project and of IB approach. Discussion about conduction in ohmic conductors. Search for evidence of non-ohmic behaviour, as in semiconductor devices.	2
Exploration	Students acquired information and planned their activities in small groups, trying to pose questions they would answer during the experimental activities. They were introduced to the laboratory and encouraged to explore the measurement facilities and materials available, in order to design their own experiences. Students chose to address the electrical conduction process in vacuum tubes, which is easier to discuss and shows marked non-ohmic behaviour.	3
Explanation	Students carried out their research investigations, designed on the basis of the hypotheses and questions formulated during the explorative phase. They decided to study the anodic current vs. the filament temperature, to collect information about the values of concentration of electrons emerging from the filament. Mathematical modelling procedures were discussed in order to find a law to describe the concentration vs. temperature trend, which was found to contain the general BF expression. Some students searched for suitable models to make sense of their experimental evidence and the specific form of the suitable function they found, in particular with respect to the meaning of the quantity "energy" contained in the law's exponential term. Some time was devoted to the analysis of an agent-based computer model related to the subject, built by using the NetLogo simulation environment (http://ccl.northwestern.edu/netlogo/), which can easily simulate the interactions between a large number of elements. Students discussed a simulated mechanical model of a two-level system with the instructor. Particularly, they dealt with a large number of balls free to move on two connected planes, placed at different heights. Using the NetLogo simulation, it was possible to study the equilibrium distribution of the balls at the two levels and discuss the factors that influence this distribution. Finally, students compared simulation findings, experimental results and models explaining them.	10
Elaboration	Students searched for physical and chemical situations different from the ones discussed during the previous activities, whose experimental dependence on temperature gives evidence of a similarity with electrical conduction in semiconductors and thermionic tubes. A final scientific report was written by each group, with students sharing their ideas and preliminary results with the other participants.	3
Evaluation	Students presented the most significant findings obtained as a result of their experimental work and held a class discussion aimed at comparing and contrasting the results obtained by different groups.	2

Table 2. *"Traditional" workshop*

Phase	Description of the activities	Hours
1	The teacher gave a lecture to students, discussing with them the pedagogical methods (laboratory and modelling activities, group work and discussions) that were to be used during the workshop and describing the contents that will be dealt. Particularly, concepts regarding electric conduction in conductors and semiconductors were recalled and the dependence of resistivity on temperature was discussed.	2
2	The teacher discussed the basic topics related to electric conduction in vacuum tubes, presented as an easier situation with respect to conduction in semiconductors. The special relevance of the concept of minimum energy that electrons must have in order to participate to the conduction process in vacuum tubes and semiconductors was highlighted. In the particular case of semiconductors, the teacher discussed the energy band evidencing the role of the energy gap that electrons must cross in order to take part to the conduction process. The easier situation of conduction in a vacuum tube was presented and the related functioning mechanism was discussed in the light of Richardson law. This law was introduced because it allows the students to easily understand the relevance of a comparison between the thermal energy possessed by the electron and the threshold energy related, in this case, to electrostatic potential energy.	4
3	During this laboratory-based phase, the teacher first described the laboratory equipment, then presented to the students the experiments and the related measurements. He performed them, showing the results to the students and gave them some worksheets that they had to use in order to repeat the measurements, take note of the results and represent them in table and graphic form. The students measured the values of electric current and temperature and the concentration of electrons coming out from the vacuum tube cathode, in order to verify that these values are in accordance to the Boltzmann Factor formula.	6
4	The teacher showed to the students a NetLogo simulation, that is a mechanical example of a two-level system related to the Boltzmann Factor. The teacher made clear that this example can be related to the experimental situation they analyzed in the previous phase and allowed the students to run the simulation, modifying some parameters in order to obtain results that agree with their experimental values.	4
5	Students worked on their experimental and simulation results by following some worksheets given by the teacher and prepared a final scientific report to discuss with the teacher.	4

3. Data analysis

The quantitative analysis methods used in this study are based on clustering techniques. They allow us to subdivide the students in groups on the basis of their typical ways (or "strategies") to tackle the questionnaire. Cluster Analysis (*ClA*) (Everitt *et al.* 2011) aims at classifying subject behaviours in different groups, or clusters. These can be analysed in order to deduct their distinctive characteristics and to point out similarities and differences between them. Here we will only use a specific not-hierarchical clustering method, called *k-means* (MacQueen 1967), as it allows the researcher to clearly individuate clusters that are also easily represented in Cartesian graphical form.

IOP Conf. Series: Journal of Physics: Conf. Series **1076** (2018) 012019 doi:10.1088/1742-6596/1076/1/012019

Student answer classification

In order to apply *ClA* method it is necessary to categorise the questionnaire answers and code them. Due to the open-ended nature of the questions, after the questionnaires were submitted to the student samples the researchers independently read the students' answers in order to empirically identify the main characteristics of the different student records (the raw data). They agreed to independently construct a coding scheme by means of a phenomenographic approach (Marton & Booth 1997) to the student answer analysis, and through the identification of keywords that were relevant for the understanding of these records. During a first meeting, the selected keywords were compared and contrasted, and then grouped into categories based on epistemological and linguistic similarities that are actually the typical answering strategies deployed by the students when tackling with the questions. As a third step, each researcher read the student records again and applied the new coding scheme, by assigning each student to a given category for each question.

At the end of the coding procedures, two shared list of M answering strategies to be used for the subsequent analyses was obtained. More specifically, $M_{pre} = 53$ answering strategies were obtained for the pre-instruction test analysis and $M_{post} = 55$ ones were obtained for the post-instruction test analysis. Each of the N = 36 students in each group was identified by two arrays, a_i and a'_i (i = 1, 2, ..., N) composed by M_{pre} and M_{post} components 1 and 0, respectively. In each of these arrays, 1 was assigned when the related student used a given answering strategy to respond to a question, and 0 when he/she did not use it. More detail of the procedure we followed to analyze data can be found in Battaglia, Di Paola & Fazio (2017).

The k-means method

Non-hierarchical clustering analysis is used to generate groupings of a sample of elements (in our case, students) by partitioning it and producing a smaller set of non-overlapping clusters with not hierarchical relationships between them. Like other clustering algorithms, k-means requires that an index to measure the likeness between two elements be defined. In the case of k-means a metric must be defined to give a measure of the likeness between two elements (the students) by using the distance deduced by the metric itself. The distance can be defined as a function of the correlation coefficient between student couples (Di Paola *et al.* 2016, Battaglia *et al.* 2017).

The results of the k-means algorithm are plotted in a 2-dimensional Cartesian space where the points represent the students of the sample, placed in the space according to their mutual distances. As we said before, for each student, we know the distances between him/her and all the other students of the sample. It is, then, necessary to define a procedure to find two Cartesian coordinates for each student, starting from these distances. This procedure consists in a linear transformation between a N-dimensional vector space and a 2-dimensional one and it is well known in the specialized literature as *multidimensional scaling* (Borg & Groenen 1997). For this reason, the X- and Y-axes simply report the values needed to place the points according to their mutual distance and are reported in arbitrary unit. Then, the k-means algorithm starts by choosing the number, q, of clusters one wants to populate and of an equal number of "seed points", initially randomly placed in the same bi-dimensional Cartesian space where data are represented. The students are then grouped on the basis of the minimum distance between them and the seed points. Starting from an initial classification, students are iteratively swapped from one cluster to another. The students belonging to a given cluster are used to find a new point, representing the average position of their spatial distribution. This is done for each cluster Cl_k (k = 1, 2, ..., q) and the resulting points are called the cluster *centroids* C_k (Leisch 2006). This process is repeated and ends when the new centroids coincide with the old ones.

In order to define the number q of clusters that best partitions the sample on the basis of the student distances it is possible to use several methods. Here we used the so-called Silhouette function (Rouseeuw 1987). For each selected number of clusters, q, and for each sample student, i, assigned to a cluster k, with k = 1, 2, ..., q, a value of the Silhouette Function $S_i(q)$ is calculated.

$S_i(q)$ gives a measure of how similar student i is to the other students in its own cluster, when compared to students in other clusters. It ranges from –1 to +1: a value near +1 indicates that student i is well-matched to its own cluster, and poorly-matched to neighboring clusters. If most students have a high silhouette value, then the clustering solution is appropriate. If many students have a low or negative silhouette value, then the clustering solution could have either too many or too few clusters (i.e. the chosen number, q, of clusters should be modified).

A remarkable feature of the centroid array is that it is composed by the answering strategies most frequently given by students belonging to each cluster (Battaglia *et al.* 2017). So, once the appropriate partition of data has been found, we decide to characterize each cluster of students in terms of its centroid.

4. Results

All the clustering calculations on the student answers to the pre- and post-instruction questionnaires were performed using a custom software, written in C language. The graphical representations of clusters in both cases were obtained using the well-known MATLAB SOFTWARE (2015). By using the Silhouette method, we found that the best partitions of our samples are achieved by choosing three clusters in both the pre- and post-tests for the experimental and the control groups.

Fig. 1a and 1b show the representations of these partitions in 2-dimensional graphs for the pre-instruction test of experimental (a) and control (b) groups, respectively. The clusters show the partition of the samples into groups made up of different numbers of students. As said above, the clusters $Cl_{k,exp}$ and $Cl_{k,con}$ (k = 1, 2, 3) can be characterized by their related centroids, $C_{k,exp}$ and $C_{k,con}$, respectively. The same will be done for the clusters we find in the post-test analysis. The answering strategies most frequently applied by students in the clusters are synthetically resumed in tables 3 and 4.

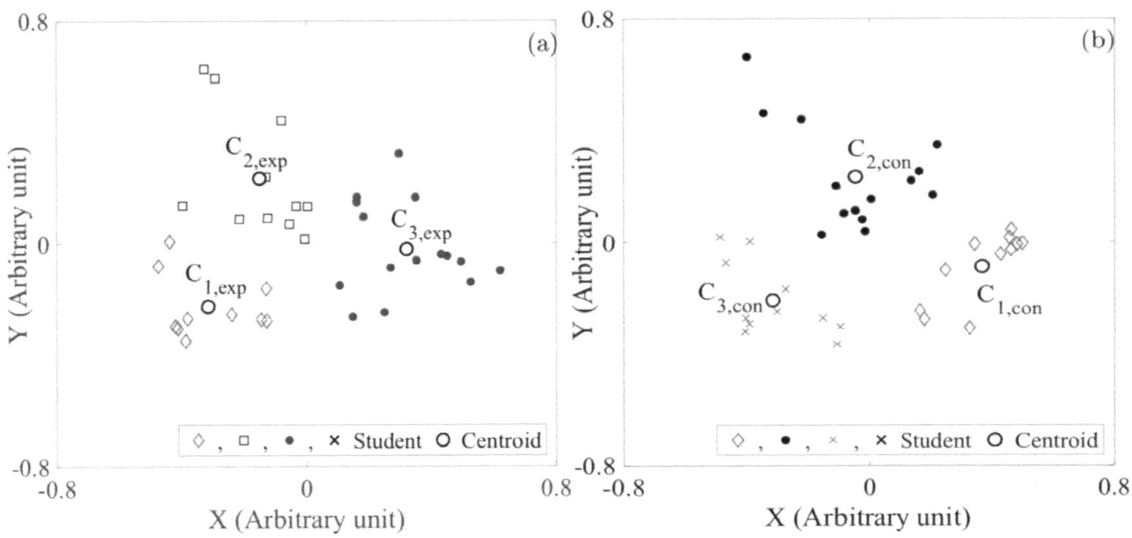

Fig. 1. k-means graphs for pre-instruction test of the experimental group (a) and control group (b). Each point in this Cartesian plane represents a student. Points labeled $C_{1,exp}$, $C_{2,exp}$, $C_{3,exp}$, $C_{1,con}$, $C_{2,con}$, $C_{3,con}$ are the cluster centroids

Table 3. An overview of results obtained in the experimental group pre-test

Cluster centroid	$C_{1,exp}$	$C_{2,exp}$	$C_{3,exp}$
Most frequently given answers	1) Only macroscopic description. 2) No description of the meaning of quantities. 3) A catalyst acts speeding-up a chem. reaction. 4) Arrhenius law is only mathematically described. 5) only a few phenomena from real-life and not relevant are mentioned. 6) Some similarities are found, *E* or *T* are mentioned.	1) verbal description and rough microscopic model given. 2) Some description based on real-life experiences. 3) A catalyst acts speeding-up a chem. reaction. A reference to energy gap is given. 4) Arrhenius law is only mathematically described. 5) Some related phenomena are described. 6) Some similarities are found, *E* or *T* are mentioned.	1) Some relevant description is given. No explanation. 2) Some quantity is described. 3) A catalyst acts speeding-up a chem. reaction. A simple reference to energy is given. 4) Arrhenius law is only mathematically described. 5) Some related phenomena are described, limited to the chemical context. Some math explanation. 6) Some similarities are found, not really relevant.
No. of students	10	11	15

Table 4. An overview of results obtained in the control group pre-test

Cluster centroid	$C_{1,con}$	$C_{2,con}$	$C_{3,con}$
Most frequently given answers	1) Some relevant description is given. No explanation. 2) Some quantity is described. 3) A catalyst acts speeding-up a chem. reaction. A simple reference to energy is given. 4) Arrhenius law is only mathematically described. 5) Some related phenomena are described, limited to the chemical context. Some math explanation. 6) Some similarities are found, not really relevant.	1) verbal description and rough microscopic model given. 2) Some description based on real-life experiences. 3) A catalyst acts speeding-up a chem. reaction. A reference to energy gap is given. 4) Arrhenius law is only mathematically described. 5) Some related phenomena are described. 6) Some similarities are found, *E* or *T* are mentioned.	1) Only macroscopic description. 2) No description of the meaning of quantities. 3) A catalyst acts speeding-up a chem. reaction. 4) Arrhenius law is only mathematically described. 5) only a few phenomena from real-life and not relevant are mentioned. 6) Some similarities are found, *E* or *T* are mentioned.
No. of students	11	14	11

Figs 2a and 2b show the results of the post-instruction tests for the experimental (a) and control (b) groups, respectively. The most frequently used strategies in the two groups are synthetically resumed in tables 5 and 6.

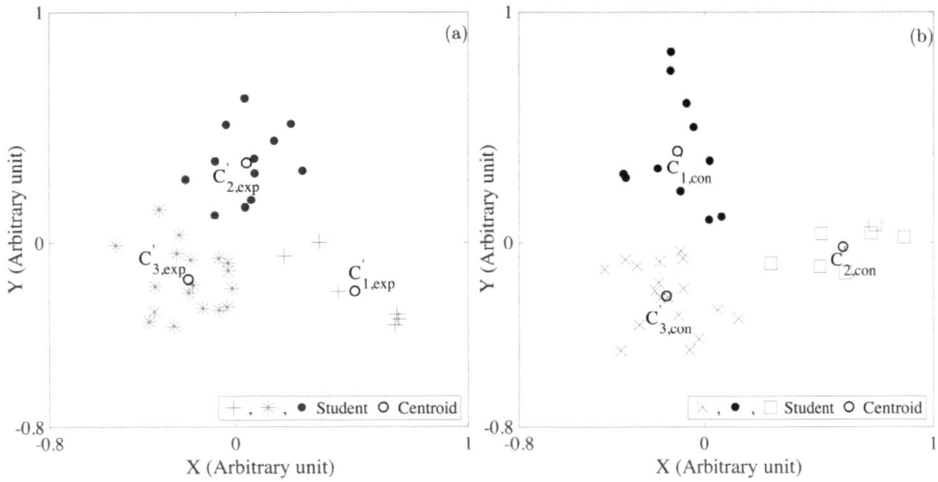

Fig. 2. k-means graphs for post-instruction test of the experimental (a) and control (b) group. Each point in this Cartesian plane represents a student. Points labelled $C'_{1,exp}$, $C'_{2,exp}$, $C'_{3,exp}$, $C'_{1,con}$, $C'_{2,con}$, $C'_{3,con}$ are the cluster centroids

Table 5. An overview of results obtained in the post-instruction test of the experimental group

Cluster centroid	$C'_{1,exp}$	$C'_{2,exp}$	$C'_{3,exp}$
More frequently given answers	1) the situation described and rough microscopic mechanism given. 2) relevant quantities found and physically described. 3) Additives are discussed in terms of energy gap. 4) Explanation in terms of molecular interaction is given. 5) some relevant phenomena are given, also not chemistry related. 6) Similarities are found. Activation energy is correctly discussed.	1) the situation described and rough microscopic mechanism given. 2) relevant quantities found and physically described. 3) Additives are described as speeding-up the flow. A rough reference to energy gap is given. 4) A rough explanation based on interaction between molecules is given. 5) some relevant phenomena are given, but only chemistry related. 6) Similarities are found. *E* and *T* are mentioned.	1) the situation is mathematically described. 2) relevant quantities found and physically described. 3) Additives are chemically described, in macroscopic terms. 4) The law is physically outlined. 5) some relevant phenomena are given, also not chemistry related. 6) Similarities are found. *E* and *T* are mentioned.
No. of students	6	12	18

Table 6. An overview of results obtained in the post-instruction test of the control group

Cluster centroid	$C'_{1,con}$	$C'_{2,con}$	$C'_{3,con}$
More frequently given answers	1) No good description, but molecular energy is cited. 2) relevant quantities are cited. 3) Additives are described as speeding-up the flow. 4) The law is mathematically described. No explanation given. 5) Some phenomena found, only related to Chemistry. 6) no clear answer.	1) Some description is given and molecular energy is cited. 2) relevant quantities are physically described. 3) Additives are described in macroscopic terms. 4) A rough microscopic model is given. 5) Some phenomena found, only related to Chemistry. Some explanation. 6) Some similarity found, *E* and *T* are found relevant.	1) Some description and a rough microscopic model are given. 2) relevant quantities are cited. 3) Additives are considered catalysts. 4) The law is mathematically described. 5) Some phenomena found, only related to Chemistry. 6) Some similarity found, but physical quantities not relevant are cited.
No. of students	11	7	18

5. Discussion and conclusion

The interpretation of *ClA* results mainly involves the identification of the typical features characterizing students' answers belonging to the same cluster as well as differences and similarities in answering strategies of students belonging to different clusters. On the basis of the obtained results, we can discuss the answer to our research question.

First of all, we note that similar behaviours can be detected in both the experimental and control group pre-test results, as the strategies in centroid arrays $\overline{a}_{k,exp}$ and $\overline{a}_{k,con}$ are almost identical. This is an expected result, as the experimental and control groups are random partitions of a unique set of students, and so should be roughly equivalent. The results show that both the students that attended the Inquiry-Based workshop (experimental group) and the ones that were exposed to a more traditional one (control group), before instruction mainly highlight the use of lines of reasoning in many cases not well suited to the study of physics. In fact, students of both the experimental and the control group seem to often use answering strategies which are inefficient to correctly find a microscopic functioning mechanism and to build proper explanations on the basis of the variables found as relevant for a phenomenon. In some cases ($Cl_{2,exp}$), the phenomenon is explained in terms of a rough functioning mechanism. Very often, reference to a well-known mathematical model seems to stimulate a recalling procedure, i.e. a search in memory for real-life examples or studied concepts) that fit in with the formula, in some cases without a clear understanding of its physical meaning. Arrhenius law is always described in pure mathematical form, without a reference to its physical meaning. Finally, in many cases students highlight a lack of generalization skills, being limited in their answers to questions 5 and 6 to the context of studied subjects. However, in some cases (clusters $Cl_{2,exp}$ and $Cl_{2,con}$) a search for a common microscopic model for the situations recalled in answers to the last two question is present. All in all, we spot here a significant use of approaches based on common-type knowledge, even if in some cases in conjunction with higher level, descriptive strategies based on previous study or with a search for rough functioning mechanisms.

The results of the analysis of student answers to the post-instruction test, however, show that a difference between the experimental and the control group students, can be identified. In fact, 6 of the 36 students in the experimental group (cluster $Cl'_{1,exp}$) are able to explain the situations and problems proposed in the questionnaire relating them to a functioning mechanisms based on the idea of thermal activation. Moreover, the students grouped in clusters $Cl'_{2,exp}$ and $Cl'_{3,exp}$, although in some cases still anchored to memories of past studies, showed to be able to explain the flow process in mathematical terms or by citing a functioning mechanism. They discuss the role of an additive by considering the energy gap concept but frequently do not relate it to interaction between molecules. However, in some cases the Arrhenius-like expression for viscosity is interpreted in terms of interaction between molecules. Finally, they seem to possess generalization skills, even if in some cases limited to familiar contexts.

Many of the students of the experimental group seem to have developed lines of reasoning about the Arrhenius-like phenomena that help them to make explanations coinciding with those of the accepted physical model and correct predictions of the behaviour of proposed situations perceived as similar. Moreover, the recognition of the common mathematical form in Arrhenius-like laws is in many cases linked to a better understanding of the functioning mechanisms behind these laws, something that was present at a substantially lower level in the initial phases of the workshop. In many cases, before the workshop activities students put the mathematical description in first place in their lines of reasoning. After the inquiry-based workshop activities, where they were encouraged to search for answers to proposed situations and phenomena by performing measurements and building models in a peer-to-peer set-up, many students appear to look at microscopic models that can explain the experimental evidence first, and then discuss and make sense of the mathematical law common form.

Some of the students of the control group also show a general improvement in reasoning with respect to the one highlighted in the pre-instruction results. In fact, students in $Cl'_{2,con}$ are able to correctly find and physically interpret the variables relevant in Arrhenius law, to discuss the role of additives in terms of the energy gap concept (although only at macroscopic level), to give an explanation of the flow processed in terms of interaction between molecules, and to find and discuss phenomena that are considered similar to the proposed one. On the other hand, the majority of the other students mainly base their approaches on a reasoning based on memory of past studied subjects or on macroscopic or mathematical explanation, without clear reference to the search for a microscopic functioning mechanism.

Our research shows that before the Workshops many of the Engineering undergraduates demonstrated mixed abilities with respect to the modelling of phenomena, initially perceived as different, but all analyzable in the common framework of the BF. Many students clearly highlighted the use of mixed-type reasoning strategies, with particular reference to ones that appear inefficient for building explications of the observed/proposed situations. These results are consistent with data from the literature (Hrepic *et al.* 2005, Bao & Redish 2006, Corpuz & Rebello 2011), that show that the lines of reasoning students deploy in creating explanations for proposed situations or observed phenomena can be eclectic and sometimes contradictory.

The results of a Bounded/Open inquiry-based workshop, focused on Boltzmann Factor based phenomena and using tools able to stimulate experimental analysis, as well as modelling at micro level, seem to highlight the efficacy of such an Inquiry approach in developing and improving the students' lines of reasoning. They appear to be redirected to the construction of mechanisms of functioning and to the identification of common aspects in apparently different phenomena, which may be an indication of a general framework in which phenomena explication can develop. A significant side effect of the IB workshop activities is a modification of the initial propensity of students to analyze a proposed situation by first taking into account the mathematical formulas and then trying to give a physical meaning to them. After the IB instruction many students seem to demonstrate the ability to discuss a proposed physical phenomenon by using a physical model that they perceive can be used to describe and explain the phenomenology. This often happens without having to refer to the mathematical formalism first, something that Vosniadou (1994) and Greca & Moreira (2002) pointed out as a key point to highlight a real comprehension in a particular field of physics.

On the other hand, the students that attended the more "traditional" workshop, still based on a laboratory and modelling approach focused on Boltzmann Factor based phenomena, but not Inquiry Based, after instruction showed improvement in their reasoning strategies. Some of them were able to correctly find and physically interpret the variables relevant in the proposed phenomena, to correctly discuss the meaning of the energy gap and to give an explanation of the flow processed in terms of interaction between molecules. However, in many cases the majority of students involved in this workshop still highlighted forms of reasoning based on a search for analogy with studied subjects and laws and on a primary use of mathematics to make sense of the proposed situations.

We can conclude that both workshops were effective in improving the student reasoning strategies. However the IB learning environment, with the freedom it offers to the students to raise their own questions, to plan research and find different resources to answer the questions, seems to have been more effective in promoting cognitive skills oriented to explication of mechanisms of functioning than the more traditional one.

References

American Association for the Advancement of Science, AAAS (1993) *Benchmarks for Science Literacy*, Oxford: Oxford University Press.

Bao L. & Redish E.F. (2006) Model Analysis: Representing and Assessing the Dynamics of Student Learning *Physical Review Special Topics Physics Education Research* **2**, 010103.

Battaglia O.R., Di Paola B. & Fazio C. (2017) A quantitative analysis of Educational Data through the Comparison between Hierarchical and Not-Hierarchical Clustering, *EURASIA Journal of Mathematics Science and Technology Education* **13**(8): 4491-4512.

Boltzmann L. (1909) Bermerkungen über einige Probleme der mechanische Wärmetheorie, Wiener Berichte **75**, 62-100. In L. Boltzmann Wissenschaftliche Abhandlungen, **II** (Hasenöhrl F., ed.), Leipzig Barth, 1909, reissued New York Chelsea, 1969, paper 39.

Boltzmann L. (1909) Über die beziehung dem zweiten Haubtsatze der mechanischen Wärmetheorie und der Wahrscheinlichkeitsrechnung respektive den Sätzen über das Wärmegleichgewicht, Wiener Berichte 76, 373-435. In L. Boltzmann Wissenschaftliche Abhandlungen, **II**, (Hasenöhrl F., ed.), Leipzig Barth, 1909, reissued New York Chelsea, 1969, paper 42.

Borg I. & Groenen P. (1997) Modern multidimensional scaling. New York: Springer Verlag.

Clough E.E. & Driver R. (1986) A Study of Consistency in the Use of Students' Conceptual Frameworks across Different Task Contexts. *Science Education* **70**(4), 473-496.

Corpuz E.D. & Rebello N.S. (2011) Investigating students' mental models and knowledge construction of microscopic friction. I. Implications for curriculum design and development. *PPhysical Review Special Topics Physics Education Research* **7**, 020102.

Di Paola B., Battaglia O.R. & Fazio C. (2016) Non-Hierarchical Clustering to Analyse an Open-Ended Questionnaire on Algebraic Thinking. *South African Journal of Education* **36**, 1-13.

Duit R. & Glynn S. (1996) Mental modelling. In *Research in Science Education in Europe* (Welford G., Osborne J. & Scott P., eds.) London: Falmer Press, pp. 166-176.

Etkina E. (2014) Helping our students learn physics and think like scientists. In *Teaching and Learning Physics Today: Challenges? Benefits?* (Kaminski W. & Michelini M., eds.) Udine, Italy: Lithostampa, p. 63.

Everitt B.S., Landau S., Leese M. & Stahl D. (2011) *Cluster Analysis*. Chichester: John Wiley & Sons Ltd.

Fazio C., Battaglia O.R. & Di Paola B. (2013) Investigating the quality of mental models deployed by undergraduate engineering students in creating explanations: the case of thermally activated phenomena. *Physical Review Special Topics Physics Education Research* **9**, 020101.

Feynman R.P., Leighton R.B. & Sands M. (1963) *The Feynman Lectures on Physics* **I**, 42.1-42.11. Reading, MA: Addison-Wesley.

Greca I.M. & Moreira M.A. (2000) Mental models, conceptual models, and modeling. *Journal of Research in Science Teaching* **22**(1), 1-11.

Greca I.M. & Moreira M.A. (2002) Mental, physical, and mathematical models in the teaching and learning of physics, *Science Education* **86**, 106-121.

Hrepic Z., Zollman D.A. & Rebello N.S. (2005) Eliciting and Representing Hybrid Mental Models, *Proceedings of the NARST 2005 Annual Meeting*, Dallas, TX.

Johnson-Laird P.N. (2006) *How We Reason*. Oxford, UK: Oxford University Press.

Leisch F. (2006) A Toolbox for K-Centroids Cluster Analysis, *Computational Statistics and Data Analysis* **51**(2), 526-544.

MacQueen J. (1967) Some methods for classification and analysis of multivariate observations. In *Proceedings 5th Berkeley Symposium on Mathematical Statistics and Probability 1965/66*, **I** (LeCam L.M. & Neyman J., eds.) Berkely: Univ. of California Press, pp. 281-297.

Marton F. & Booth S. (1997) *Learning and awareness*. Mahwah, NJ: Lawrence Erlbaum Associates.

MATLAB version 8.6 (2015). Natick, Massachusetts: The MathWorks Inc. www.mathworks.com/products/matlab/.

Myers A. & Hansen C. (2012) *Experimental psychology*. Belmont, CA: Cengage Learning.

National Research Council (NRC) (2012) *A Framework for K-12 Science Education: Practices, Crosscutting Concepts, and Core Ideas*. Washington DC: The National Academies Press.

Pizzolato N., Fazio C., Sperandeo Mineo R.M. & Persano Adorno D. (2014) Open-inquiry driven overcoming of epistemological difficulties in engineering undergraduates: A case study in the context of thermal science, *Physical Review ST Physics Education Research* **10**, 010107.

Rouseeuw P.J. (1987) Silhouette's: a graphical aid to the interpretation and validation of cluster analysis. *Journal of Computational and Applied Mathematics* **20**, 53-65.

Sadeh I. & Zion M. (2009) The Development of Dynamic Inquiry Performances within an Open Inquiry Setting: A Comparison to Guided Inquiry Setting. *Journal of Research in Science Teaching* **46**(10) 1137-1160.

Sathya R. & Abraham A. (2013) Comparison of Supervised and Unsupervised Learning Algorithms for Pattern Classification, *International Journal of Advanced Research in Artificial Intelligence*, **2**(2), 34-38.

Vosniadou S. (1994) Capturing and modeling the process of conceptual change, *Learning and Instruction - Journal* **4**, 45-69.

Wenning C.J. (2005) Levels of Inquiry: Hierarchies of pedagogical practices and inquiry processes. *Journal of Physics Teacher Education Online* **2**, 3-12.

IOP Conf. Series: Journal of Physics: Conf. Series **1076** (2018) 012020 doi:10.1088/1742-6596/1076/1/012020

The analysis of conceptual understanding and selected inquiry skills of students within non-formal education activities in the science Centre

Mária Bilišňanská[1], Marián Kireš[2]

[1]Division of Didactics of Physics, Institute of Physics Faculty of Science P. J. Šafárik University in Košice, Slovakia
[2]Division of Didactics of Physics, Institute of Physics Faculty of Science P. J. Šafárik University in Košice, Slovakia

Abstract. Science centers have been established in an effort to promote science and technology in society, among children and young people. They are basically institutions that have prepared a series of scientific experiments, observations, games and similar attractions working on a selected phenomenon. A surprising effect, unexpected behaviors or inexplicable phenomena that are supposed to raise curiosity and desire to discover in a visitor are offered to them as a rule. Visitors of science centers are often groups of school children on a school excursion. From a teachers' point of view it would surely be welcome if besides motivation the excursion had an educational aspect and if it could be possible to use it to complement subject matter or to develop selected skills, which schools do not have appropriate conditions for. For support of such ideas we established Inquiry science laboratory within the science center SteelPARK in Košice, where groups of school students can participate on two different inquiry activities prepared for each mostly one month period. Activities are guided by a lecturer, a pre-service physics teacher, in parallel for 5 groups of 3-4 students. Our laboratory teaching centers around key problems, pre-selected for each inquiry activity. Total time for one activity is 60 minutes. Conceptual test questions are discussed during the first part of the activity as a tool for the student's involvement into the problem and also as a source of motivation. Activities are on the level of guided inquiry, which we specified through examples of ESTABLISH project units. After the activity, formative assessment tools are used for self-evaluation of selected development skills. In the paper we present the results of the conceptual tests and rubrics, basically created within the SAILS project, for skills mapping across the sample of more than 600 students discussed and analyzed on the basis of one exemplary activity. After two years of experience with more than 4500 participants from lower and upper secondary schools, the level of selected inquiry skills across 15 different inquiry activities is presented. All of our worksheets as well as findings from non-formal educational activities are recommended for use by teachers in school practice.

1. Introduction

Informal learning and popularization events in science centers are used all over the world to increase the interest of youth and to lay the foundation of a positive stance on science in society. Science centers have a long tradition in many metropolitan cities, exploratory and entertainment – learning parks, such as Cité des Sciences et de l'industrie in Paris, Universum in Göteborg, Science Museum in London, CSOPA in Budapest, etc. Attendees learn about the laws of nature, of The Earth and The Universe, through interactive experiments and games. The interest of youths in their surroundings is am-

IOP Conf. Series: Journal of Physics: Conf. Series **1076** (2018) 012020 doi:10.1088/1742-6596/1076/1/012020

plified through the use of entertaining experiments; they discover the fantastic world of physics and other natural science disciplines through unforgettable experiences. Nowadays, when real life experience are being more and more supplemented by computer simulation, students can try by hands, how real experiments works. Each of the science centers is specific in some way; it presents its own story and approach to its visitors. The same is true about the new science centre SteelPARK Košice (www.steelpark.sk), the creation of which we took active part in. The visitors can find out about the history of steel – from its mining, refining and its final products. More than 50 interactive exhibits present the know-how and creativity of several scientific disciplines connected with the manufacture and use of products of metallurgy.

After two years of service, we have found that a large portion of visitors consists of school students. There was a natural request, from the teachers, when planning repeat visits to the center, to utilize the motivational function of the science center to strengthen the educational impact, mainly in the area of developing specific research capabilities of students of primary and secondary schools. Because of this we started the Inquiry science lab in October of 2014.

Inquiry science lab in the science center

Inquiry science lab (ISL) has been created for students of primary and secondary schools and their teachers, so that they can acquire their own experience with inquiry based education, through the use of non-formal learning. The activities are, from the view of classical education, untraditional measurements, and observations connected with the discovery of new findings and mainly acquiring inquiry skills. The students work in three or four person teams, under the guidance of a lecturer and using worksheets. The teacher is in the role of and observer, observing the progress of IBSE (Inquiry Based Science Education), which is on the level of guided inquiry (SAILS project). The activity in the ISL takes 60 minutes, and 2 different activities are run in tandem, in order to be able to include all the students of an entire class.

Fig. 1. Inquiry science lab in SteelPARK

The learning activities, worksheets, methods and tools of evaluation are available to the teachers and can use them later in their education in schools. ISL is also used to support the practical education of pre-service teachers of interdisciplinary study in combination with physics. Through lecturing they are preparing for their future career in teaching. They acquire important experience with inquiry based activities, they are thought to put them into practice, take part in the collection of data, and are informed on the results of didactic surveys.

During the 18 months of existence of ISL, 6383 visitors took part in 15 different activities. The annual attendance rises every year. In two academic years (2014/2015, 2015/2016) 15 activities were conducted, on which around 400 groups of students took part. Table 1 shows the number of students who took part in the individual activities.

IOP Conf. Series: Journal of Physics: Conf. Series **1076** (2018) 012020 doi:10.1088/1742-6596/1076/1/012020

Table 1. Activities of ISL in the academic years 2014/2015 and 2015/2016

Activity	Number of students PS	Number of students SS	Together
Academic year 2014/2015			
1. Can we measure the weight of air?	134	183	317
2. What can we learn from a laser range finder?	156	357	513
3. Battle on an unstable basin	210	26	236
4. Become a forensic scientist	342	81	423
5. How do we breath?	295	113	408
6. Fill it up with water and go	298	151	449
7. How does a bat see motion?	420	185	605
Total			**2951**
Academic year 2015/2016			
8. What does a scale show?	410	102	512
9. How does a candle burn?	410	102	512
10. YO-YO, child's play full of physics	71	82	153
11. On the heels of a physics scale	220	71	291
12. Where do we get the air we breathe?	350	185	535
13. Known and unknown shadows	387	163	550
14. Let us repair Galileo's thermometer	406	101	507
15. How to take pictures of moving objects?	298	72	372
Total			**3432**

2. The mapping of students' conceptual understanding by use of pre-test

We have conducted a didactical research on the preconceptions of students, the development of specific inquiry capabilities and elements of scientific knowledge on a wide range of visitors within the individual activities. At the beginning of the activity, the attendants take and entry conceptual test on keywords connected with the activity, which is supposed to motivate active learning.

We would like to show the concepts researched in the activities by illustrating it on an example "YO-YO, child's play full of physics", which is concerned with the speed of the falling yo-yo. 149 students took part, choosing an answer from 5 different options.

Q1 The speed of a downward motion of a yo-yo:

a) Uniformly increases.
b) Stays constant.
c) Firstly it stays constant, then it decreases with the final stop at the lowest position.
d) Firstly it increases and then it decreases with the final stop at the lowest position.
e) Uniformly decreases.

A staggering 65% of the students have the wrong idea about the yo-yos downward motion.

When the yo-yo reaches its lowest point, we feel a slight jolt. This is caused by the change in the direction of its speed. The students had to choose from 4 different options.

Q2 When the yoyo reaches the lowest position we can feel the slight force acting on our hand. The reason is:

a) We do it ourselves in order to make yoyo to continue in an upward motion.
b) The change in direction of rotation.
c) Change in the direction of velocity (from descending to rising motion).
d) The string that is not ideally smooth and the energy losses.

86% of students of secondary and 96% of primary schools failed to answer correctly.

Q3 At the lowest position, the rotational motion of a yo-yo:

a) Stops and then it starts in the same direction.
b) Stops and then it starts in the opposite direction.
c) Does not stop, it continues in the same direction.
d) Does not stop, it continues in the opposite direction.

Q4 After reaching the lowest point, the yo-yo can move up because:

a) We tug upward with a string by a hand and this way we supply energy to motion (Without tugging the yo-yo would not move upwards).
b) The yo-yo has kinetic energy that can be changed into potential energy.
c) The string works as a spring that is stretched.

When comparing the answers of primary and secondary school students, we have concluded, that there are no relevant differences in conceptual understanding of the given theme (see Fig. 2).

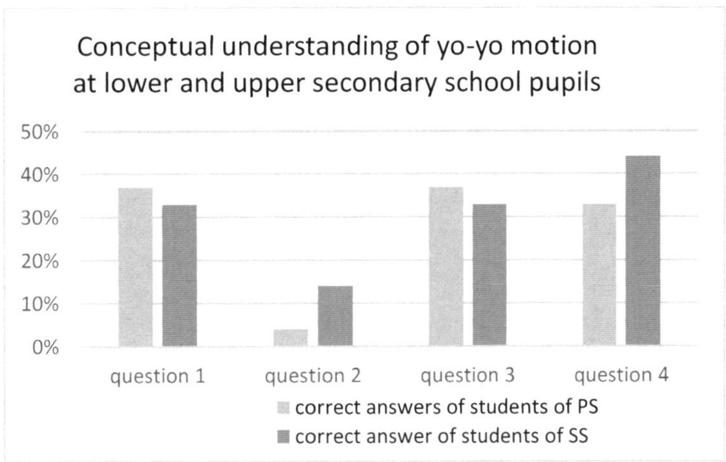

Fig. 2. Correct answers of students of PS and SS on individual questions on the pre-test

In the activity "How does a candle burn?" which 512 students took part in, we mapped the students predictions about the temperature of the flame.

Q Imagine a candle flame. Think first and then answer:

a. What is the flame temperature?
   Write down the temperature and describe how you have decided about this value.
b. Is the candle flame temperature equal all over of the flame?
   Explain your answer.

69% of the students got the answer wrong, it was less than 300°C), and the factors influencing the time of burn, whereas only 24% of the students answered satisfactorily.

At the beginning of the activity "Can we measure air?" we gave 144 students a question without an exact answer.

Q How much air weights in a room with the volume of $100m^3$.

13% of those who took part said that air weighed nothing, 39% gave a very low number, mostly in grams, 44% did not answer at all (Fig. 3). Only 4% wrote the correct relationship for calculating the weight of air, which was not given because of the absence of knowledge of the air density.

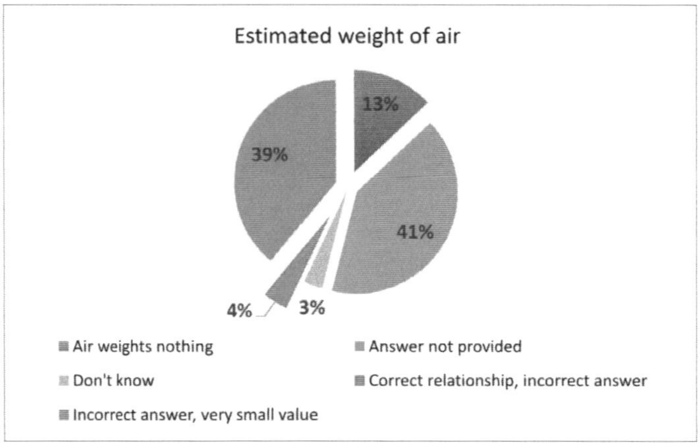

Fig. 3. Students preconceptions about the weight of air in a room

Guided inquiry with the help of worksheets

The introductory motivation on the given theme in the form of a pre-test and the evaluation of correct answers directly precedes the activities in the worksheet. The students act in accordance with the manual, with pre-defined steps. They acquire the skills of scientific work by way of observation, evaluation, measurements, collecting and evaluating data (Wenning 2007). They have an opportunity to write down their predictions on the worksheet, to plan their experiment, argue and make conclusions (Van den Berg 2013). The worksheets are scanned and stored in pdf, after the activity, for future reference. By analysis the worksheets and also by observing the work of the students during the activity, we can see reasonable reserves of the students in the following capabilities of scientific work:

- problems with writing own prediction, remarks, comments (not given by teacher),
- to argue,
- to plan an experiment,
- work as a team, discuss topics, personal responsibility for team work,
- reading with understanding.

Personal feedback and self-evaluation

At the end of the activity the students, by means of a self-evaluating sheet, evaluate the degree of their own abilities (through rubrics with three grades), which were developed. They also assess what they have learned, what they found most interesting, what they still do not understand. Through the use of self-evaluation rubrics we focus on the formative evaluation of the students work.

During the activity "What can we learn from a laser range finder?" (Ochoa *et al.* 2014), the students researched the refractive index of water, based on the fraction of the length of the aquarium measured above and below the level of the water. They recorded their measurements in a table. After completing the activity the students evaluated four capabilities developed, through rubrics with three grades (complete the activity with strong support, with support, independently) (Hattie & Timperley 2007).

Table 2. Self-evaluation rubrics, filled out by students after the activity

Selected skills *"I'm able to"*	With strong support %	With support %	Independently %
use a laser range finder	2	13	85
save data into a table	4	16	80
determine refractive index of liquid	5	35	60
explain meanings of refractive index	12	50	38

With activities with manual focus, we can achieve satisfactory results in the development of skills. The amount of development of skills drops with activities which are more intellectually challenging. It will be important to strengthen the activity from the point of understanding and the ability to explain the meaning of the refractory index.

507 students took part in the “Let us repair Galileo’s thermometer”. It seems at first glance as an easy explanation for the principles of a thermometer through the use of Archimedes’ law. In reconstructing the thermometer, the original liquid is replaced with water, which has a higher density. The students are faced with a problem, to use their current knowledge in solving a new challenge. Guided inquiry teaches the students about the importance of acquiring new experimental data and of its evaluation. We have observed, during the activity, some persisting problems with:

- making conclusions with help of own data,
- using knowledge for solving a new problem.

We intend to share the acquired results with teachers, to show them the importance of practical understanding of Archimedes’ law and the missing skills of applying their knowledge to new problems.

3. Conclusions and recommendations

Guided inquiry activities are well accepted by students because they are always looking for new information. There are no relevant differences in the conceptual understanding of selected concepts between primary and secondary school levels – secondary schools have to focus stronger on it. In some skills, the level of upper secondary school students is lower than lower secondary one. It is our opinion that this is caused by the fact that secondary school education is dominated by a theoretical form of knowledge being educated and by the methods of interpretation of teachers. Our challenge toward the teachers is: Next, let’s try activities from Inquiry science lab with your students at your school!

Acknowledgements

This article was created with the help of SRDA-0715-12: Research of the effectiveness of the methods of innovation of teaching mathematics, physics and informatics and of the project Erasmus+ SciVis 2014-1-DE01-KA203-000694: Bettering interactive methods for the understanding of natural scientific and technological solutions.

References

The ESTABLISH project. Web pages available on http://www.establish-fp7.eu.

Hattie J., Timperley (2007) The Power of Feedback, *Review of Educational Research*. March 2007, **77**(1).

Ochoa R., Fiorillo R., Ochoa C. (2014) Index of refraction measurements using a laser distance meter. *Physics Education*, March 2014, **52**(4).

The SAILS project. Web pages available on http://sails-project.eu.

Van den Berg E. (2013) The PCK of Laboratory Teaching: Turning Manipulation of Equipment into Manipulation of Ideas, *Scientia in educatione* **4**(2), 74-92, available online on http://www.scied.cz/index.php/scied/article/viewFile/86/72.

Wenning C. (2007) Assessing inquiry skills as a component of scientific literacy. *Journal of Physics Teacher Education Online* **4**(2), winter 2007, pp. 21-24. Scientific Inquiry Literacy Test (ScInqLiT), available online: http://www2.phy.ilstu.edu/pte/publications/assessing_ScInq.pdf.

IOP Conf. Series: Journal of Physics: Conf. Series **1076** (2018) 012021 doi:10.1088/1742-6596/1076/1/012021

Impact of inquiry activities in physics teaching on the level of students' inquiry skills

Brigita Balogová, Zuzana Ješková

Pavol Jozef Šafárik University in Košice, Slovakia

Abstract. Currently we face a strong effort to change science education towards shift from the traditional way of teaching and teacher-centred to student-centred education, what is the basis of Inquiry-Based Science Education (IBSE) approach.

Within the Slovak national project VEMIV aimed at research on the efficiency of innovative teaching methods in mathematics, physics and informatics education there has been a research question about the level of students' inquiry skills and the impact of systematic implementation of inquiry activities solved. In order to evaluate the level of inquiry skills a test of inquiry skills was developed. It focuses on the following inquiry skills: formulation of hypothesis, a design of an experiment (which variables, relationships), presentation of the results in a standard form (tables, graphs), determination of qualitative and quantitative relationships from graphs, tables and equations, determination of accuracy of experimental data (identification of possible sources of errors), explanation of relationships and scientific argumentation. After completing the test students from selected classes of six experimental schools experienced consistent and systematic implementation of IBSE at different levels of inquiry. The activities in physics include teaching materials in the form of students' worksheets complemented by formative assessment tools. The activities are mostly designed at interactive demonstration and confirmation or guided inquiry level. Each activity is built up in an inquiry manner, i.e. it starts with the physical problem/question that students are expected to examine. The problem is followed by formulating a hypothesis that students want to verify, then students choose variables and design an experiment. Following the instructions and answering questions, students continue step by step through conducting experiment, collecting data and their analysis, interpreting data towards drawing conclusions. The activity is complemented with formative assessment conceptual tasks connected to the activity topic as well as the self-assessment table where students assess themselves with regard to skills and knowledge they gained.

After the period of systematic implementation of IBSE the same students answered a slightly modified inquiry skill test again. The results of 300 students, who answered both pre-test and post-test have shown positive shift in the level of assessed inquiry skills as a whole, however, the shift in individual test items and tested skills differs. In the paper the results of pre-test and post-test with the focus on questions with the physical context aimed at selected inquiry skills are analysed, compared and discussed in detail.

1. Introduction

In the last 20 years the concept of inquiry has been mentioned regularly and inquiry based science education (IBSE) has been disseminated all over Europe and after the school reform in 2008 it has been progressively implemented also into the national curricular documents of the Slovak Republic. After the reform science education in Slovakia is more and more focused on movement from the traditional way of teaching and teacher-centred education to student-centred education.

The key goal of inquiry and IBSE is to involve students in the process in which students build their own "scientific" knowledge by going through a sequence of steps. These steps correspond to the stages of inquiry defined by Linn, Davis and Bell (2004). These authors define inquiry as the "*intentional process of diagnosing problems, critiquing experiments, and distinguishing alternatives, planning investigations, researching conjectures, searching for information, constructing models, debating with peers, and forming coherent arguments*".

This method of learning helps students not only to increase the scientific knowledge, but also to develop their skills and abilities to do inquiry (Harlen 2013). It is generally agreed that IBSE enables to prepare students for the future by developing and improving their skills to face problems critically and solve them. As emphasized by Trilling and Fadel (2009), the inquiry based pedagogy can contribute to the development of key skills and competences needed for student in their future career known as a 21st Century Skills.

In Slovakia, as well as at the European level much effort has been devoted to development of educational materials that are available to teachers. The open question is how these materials are actually implemented in schools and what their effect on students' learning gains is. As a result, the national project (VEMIV) has been worked out aimed at research on the efficiency of IBSE with regard to development of inquiry skills. The main goal is to prepare teaching and learning materials, educate teachers and implement inquiry activities systematically across three subjects of mathematics, physics and informatics in order to achieve synergetic effect on the development of inquiry skills. In this study we present the inquiry activities designed for physics lessons and the analysis of the impact of systematic inquiry skills implementation on the level of inquiry skills development.

2. Research methodology

The main research question concerns the effect of consistent implementation of inquiry activities across three subjects of mathematics, informatics and physics on the development of students' inquiry skills. Among several research questions emerging from the main research question, in this study we focus on the following one, in particular:

- *What is the impact of consistent implementation of inquiry activities across three subjects on the level of students' inquiry skills with focus on physics?*

The research design involves four main parts:

1. Developing and collecting a set of inquiry activities complemented with teaching and learning materials.
2. In-service teacher training.
3. Pedagogical experiment.
4. Evaluation of findings.

Teaching and learning materials

Based on the existing inquiry activities we have collected or developed a set of teaching and learning materials that correspond to the set of inquiry activities in physics. Each activity is built up in an inquiry manner, i.e. it starts with the physical problem/question that students are expected to examine. The presented problem is usually followed by formulating a hypothesis that students verify, then identifying variables and designing an experiment. Following the instructions and answering questions, students continue step by step through conducting the experiment, collecting and analysing data, interpreting data towards drawing conclusions. With regard to the level of autonomy of students the activities are more or less open. They are designed mainly in interactive demonstration or confirmation and guided inquiry manner (Banchi & Bell 2008, Wenning 2010). In table 1 the list of examples of developed inquiry activities is presented. Each of the activity was complemented with students' worksheet. The instructions in the worksheet follow the inquiry steps. Example of an inquiry activity *What happens when you press the gas?* and its corresponding worksheet with highlighted inquiry steps is presented in Fig. 1.

Table 1. A list of inquiry activities implemented at different level of inquiry

Grade	Physics activities	Type of inquiry
1.	How carts move?	interactive demonstration
1.	How human moves?	interactive demonstration
1.	How the sailboat moves?	confirmation inquiry
1.	How the cyclist speeds up/slows down?	confirmation / guided inquiry
1.	What force baseball player hits the ball?	confirmation /guided inquiry
1.	How the ice slows us down?	confirmation inquiry
1.	How friction affects motion?	bounded inquiry
1.	How…up there?	open inquiry
2.	What happens with the object submerged into the water?	interactive demonstration
2.	What happens when you compress a gas?	confirmation / guided inquiry
2.	What happens when you heat a gas?	guided inquiry
2.	Let's explore a candle flame.	guided inquiry
2.	What does the sound look like?	confirmation / guided inquiry
2.	How fast the sound travels?	confirmation / guided inquiry
2.	How do different elements behave in direct electric circuit?	guided inquiry
2.	How temperature affects resistance?	guided inquiry
2.	What is hidden in the black box?	guided inquiry

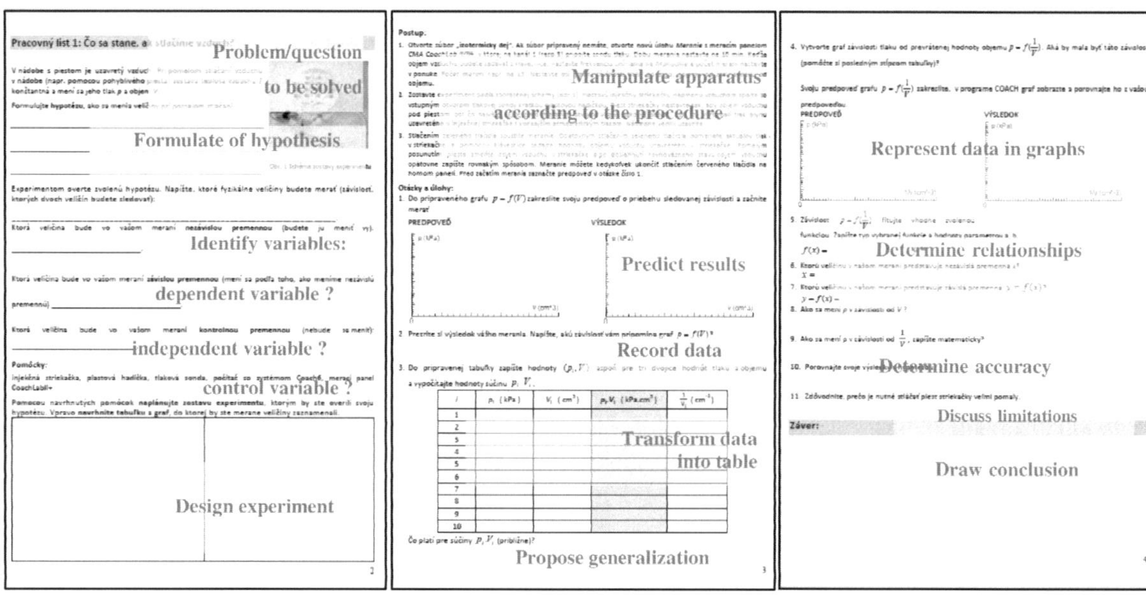

Fig. 1. Example of worksheet on the *What happens when you press the gas?* activity with highlighted inquiry steps

As reported by Harlen (2013) and Hattie (2009) formative assessment is considered an important factor to improve learning. Its goal is to monitor student learning to provide ongoing feedback. The feedback can come from teacher to students, but also from students to teacher. It helps students identify their strengths and weaknesses and target areas that need work. On the other hand, teachers recognize where students are struggling and address problems immediately. The formative assessment can be enhanced by students' self-assessment or assessment by their peers.

In accordance to these studies the worksheets were complemented with the formative assessment tasks of two types in order to enhance learning. Firstly, assignments for students were added at the end of the worksheet. It served two purposes, i.e. giving feedback to teachers as well as feedback to students about what they have learnt and how they can apply gained knowledge and skills in the situation similar to that of the conducted activity (Fig. 2).

Task 1: For pressure and volume of an ideal gas the following relationship applies under certain conditions $pV = 1550\ \text{kPa} \cdot \text{cm}^3$. Select the correct statement about the relationship between variables p and V.

a) How many times V increases, so many times *p* decreases.
b) How many times V increases, so many times *p* increases.
c) As V changes, p remains constant.
d) As p increases, V decreases.
e) As p changes, V remains constant.

Fig. 2. Example of formative assessment tasks

Self-Assessment Tool

Date: **Name:**

At today's gas activity I have learned:

The most interesting thing in this activity was:

The question, that I still cannot answer is:

EVALUate results of your work			
After the activity I can…	**with big help**	**with help**	**alone**
formulate hypotheses.			
identify dependent and independent variables.			
design experimental procedure.			
measure $p = f(V)$ relationship when the gas is slowly compressed.			
transform data into the table			
determine $p = f(V)$ relationship from the measured data.			
construct graph of $p = f(\frac{1}{V})$ relationship.			
fit $p = f(\frac{1}{V})$ relationship with appropriate function.			
understand why we fit the measured data with a function.			
compare results with the hypothesis.			

Fig. 3. Examples of self-assessment sheets

Secondly, students were involved in self-assessment so they took part in reflecting the work they did and identifying what they needed to do to improve or move forward. They filled in a self-assessment sheet where they assessed themselves with regard to skills and knowledge they gained. Students evaluated themselves in several items, e.g. how well they proceeded in certain parts of the activity or how much help they needed in conducting the activity (Fig. 3 right). In the self-evaluation sheet (Fig. 3 left) they commented what they learned during the activity, what was interesting for them and they formulated a question they still could not answer.

In-service teacher training

Two summer schools were organized in order to educate teachers towards IBSE as well as the design and implementation of pedagogical experiment. Teachers of three subjects: mathematics, physics and informatics were invited to the training. First summer school was devoted to introduction to IBSE, hierarchy of inquiry activities so that teachers get insight and deeper understanding of what inquiry is about. The school year followed by the first summer school was aimed at training inquiry activities in the classroom. The second summer school focused on the use of formative assessment to enhance IBSE and the development of instrument to diagnose the level of development of inquiry skills. In addition, the planned pedagogical experiment was introduced and all the necessary actions with regard to the pedagogical experiment were discussed and agreed. Teachers designed the plan which inquiry activities they intend to implement and they agreed on the time schedule with their colleagues from the same school.

Pedagogical experiment

In order to answer the research question the following hypotheses were formulated:

H_{01}: *There is a significant increase of the level of development of selected inquiry skills after the consistent implementation of inquiry activities across the three subjects of mathematics, physics and informatics.*

H_{02}: *There is a significant difference in the development of inquiry skills connected with physics.*

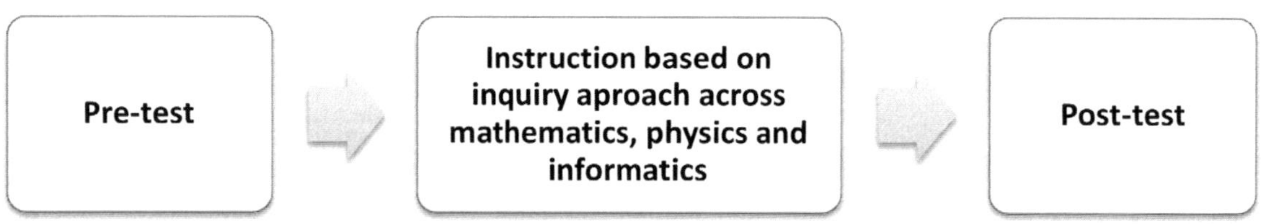

Fig. 4. Design of pedagogical experiment

The pedagogical experiment was designed as presented in Fig. 4. After completing the pre-test students were subjected to consistent implementation of inquiry activities in the period of approximately four months. During this period mathematics, physics and informatics teachers who participated in two successive teacher trainings on IBSE were implementing inquiry activities in the amount of at least three activities per subject with each activity lasting 1-2 lessons. Afterwards students took a post-test.

Sample of research

The research was carried out in the school year of 2015/2016. The research sample involved six 1st and eight 2nd grade experimental classes from six grammar schools with the total number of 368 participating students. The experimental group was subjected to the instruction based on inquiry approach across three subjects as described in the previous section.

Instrument

A test aimed at assessing development of inquiry skills was designed for the research purposes and testing hypothesis. Test items were focused on testing six inquiry skills selected from the adopted inquiry skills framework (Ješková *et al.* 2018):

- *formulate hypothesis to be tested*
- *design experiment (which variables, which relationship)*

- *transform results into standard form – graph*
- *determine relationships between variables based on: graphs, tables, test and formulas*
- *determine accuracy of experimental data (identify possible sources of errors)*
- *discuss/defend results/form arguments*

The test consists of 12 items in the context of mathematics, physics and informatics. From the total number of 12 items, 5 test items with physics context. Students could gain a score of 0-1 point for each test item, so that they got a final score in a range of 0-12 points. The test was prepared in two slightly different versions implemented as a pre-test and post-test.

3. Results and discussion

Hypotheses testing

Only those students who answered both pre-test and post-test were included into the analysis of the results. This way there were altogether 300 students involved into the analysis[1]. Table 2 represents a summary in relation to the research sample (left) and basic descriptive statistics for the test results (right).

Considering the first hypothesis testing, from the table 2 it can be seen that the mean score of post-test differs from the pre-test results shifting from 34.5% to 41.5% with the mean gain of 7%. Table 3 (left) presents the results of the paired sample t-test on comparison between pre-test and post-test showing statistically significant difference in the test score ($p < 0.001$). That means that the experimental group reached significant increase of the level of development of selected inquiry skills after the consistent implementation of inquiry activities across the three subjects of mathematics, physics and informatics (Ješková *et al.* 2016).

Table 2. General information about experimental group (left) and results of pre-test and post-test in % for experimental group (right)

Experimental group		**number of respondents**
sample size		300
grade	1st	140
	2nd	160
gender	F	158
	M	142
specialization	math/inf	76
	languages	75
	general	149

Test	**PRE**	**POST**
Mean	**34.5**	**41.5**
Median	33.3	39.4
Standard deviation	17.7	18.1
Standard error	1.0	1.1
Minimum	0.0	0.0
Maximum	92.1	90.0

Considering the first hypothesis testing, from the table 2 it can be seen that the mean score of post-test differs from the pre-test results shifting from 34.5% to 41.5% with the mean gain of 7%. Table 3 (left) presents the results of the paired sample t-test on comparison between pre-test and post-test showing statistically significant difference in the test score ($p < 0.001$). That means that the experimental group reached significant increase of the level of development of selected inquiry skills after the consistent implementation of inquiry activities across the three subjects of mathematics, physics and informatics (Ješková *et al.* 2016).

[1] The statistical analysis was carried out in R software, http://www.R-project.org

Considering the second hypothesis testing, we had a closer look on the test items put into the context of physics. If we focus on these test items, in particular (table 3, right), the mean gain in this subtest reached the mean value of 3%. The paired sample t-test proved significant improvement for this part of the test ($p = 0.033$). The mean score reached in the individual test items can be seen in table 4.

Table 3. Total mean gain for experimental group (left) and mean gain for experimental group in the subtest in the context of physics (right)

Sample	**Experimental group**
Mean gain	**7%**
Shapiro – Wilk normality test for diffrence	$p = 0.182$
Paired t-test	$p < 0.001$

Physical items 1.1, 1.2, 5, 6, 8	**Experimental group**	
Test	PRE	POST
Mean	35.7%	38.7%
Mean gain	**3.0%**	
Shapiro – Wilk normality test for difference	$p = 0.171$	
Paired t-test	$p = 0.033$	

Table 4. Mean score in % for physical items in pre and post-test

Test item	**Skill**	**PRE**	**POST**
1.1	formulate hypothesis to be tested	30.0	24.3
1.2	design experiment (which variables, which relationship)	49.4	67.3
5		36.5	44.3
6	determine relationships between variables based on graphs	33.0	38.7
8	determine accuracy of experimental data (identify possible sources of errors)	29.7	18.7

Analysis of test items in the context of physics

In the whole test five test items were related to physics. They were adapted from the existing tests of inquiry skills, such like ScInqLiT (Wenning 2007) and TIPS (Burns *et al.* 1985). These test items were aimed at testing the level of the skills typically developed in the physics lessons involving: Formulation of hypothesis (1 test item), Design experiment (which variable, which relationship, 2 test items), Determine relationships between variables based on graphs (1 test item) and Determine accuracy of experimental data (identify possible sources of errors, 1 test item). In the following sections we present selected test items connected with physics and analyse the difficulties and misconceptions identified in students' answers.

Inquiry skill *Formulate hypothesis to be tested*

Table 5. Items in pre and post-test focusing on the skill of formulating hypothesis

PRE-TEST	POST-TEST
1.1. Maria wondered if the earth and oceans are heated equally by sunlight. She decided to conduct an investigation. She filled a bucket with a kilogram of soil and another bucket with a kilogram of water. In sunny summer day she placed them so each bucket received the same amount of sunlight. Which **hypothesis** should be tested by Marie to get the answer to her question?	**1.1.** John decided to build a dog house. He thought about the question what material to use to make the house thermally insulated from the surroundings. He decided to conduct an investigation. He took two boxes of equal size made of wood and styrofoam and put a glass with the same amount of hot water into each them. Then he placed both boxes outside cools down. Which **hypothesis** should be tested by John to get the answer to his question?
a) How is water and soil heated by the sun?	a) The longer the wood and styrofoam boxes are outside, the colder the water becomes.
b) The longer the soil and water are in the sun, the warmer they become.	**b) The water in the wooden and styrofoam box cools differently.**
c) Soil and water are heated by the sun differently.	c) The wood and styrofoam box conducts different amounts of heat at different times of the day.
d) Soil and water receive different amounts of sunlight at different times of the day.	**d) The water in the wooden and styrofoam box cools equally.**
e) Soil and water are heated by the sun equally.	e) How water cools in the wooden and styrofoam box?

The test item is focused on formulation of correct hypothesis. The hypothesis should correspond to the research question: Are earth and oceans heated equally by sunlight? (in the pre-test) or What is better thermal isolator: wood or styrofoam? (in the post-test). The difficulties and misconceptions in both pre and post-test were very similar. Majority of students selected only one correct answer, even though there were two answers corresponding to correct hypothesis. It seems that once a student came across the correct hypothesis, he did not try to check the rest of possible answers. Only a small number of students selected the other correct hypothesis or both correct answers that were even less frequent in the post-test. In both pre and post-test items the most common misconception was also the confusion between the hypothesis and question. Surprisingly, in this particular test item students even did not achieve any positive shift (table 4).

Inquiry skill *Design experiment (which variables, which relationship)*

Table 6. Items in pre and post-test focusing on the skill to design experiment

PRE-TEST	POST-TEST
5. If you hang a stone on a spring scale in air or liquid, the scale shows different values. If this stone is submerged into the water, spring scale shows smaller value than in the air. The upward force exerted by the water that helps support the rock is known as the buoyant force. A group of students wants to see:	
how **mass (or weight) influences buoyant force**. They take several objects (some may be hollow).	how **volume of the body that is submerged in water influences buoyant force**. They take several objects (some may be hollow).

IOP Conf. Series: Journal of Physics: Conf. Series **1076** (2018) 012021 doi:10.1088/1742-6596/1076/1/012021

shape	**mass m (g)**	**weight G = mg (N)**	**volume V (cm^3)**	**material**	**shape**	**mass m (g)**	**weight G = mg (N)**	**volume V (cm^3)**	**material**
cube	75	0.75	27.8	aluminium	cube	75	0.75	27.8	aluminium
disk	75	0.75	53.3	copper	sphere	100	1.00	53.3	aluminium
sphere	100	1.00	53.3	aluminium	block	100	1.00	27.8	copper
cylinder	100	1.00	27.8	copper	cylinder	75	0.75	53.3	copper

Decide which two objects students should use assuming that the shape and material of the object does not play a role in the experiment.

a) Cylinder and disk, because the materials need to be the same for the experiment.
b) Cube and cylinder; because the volumes need to be the same for the experiment.
c) Cylinder and sphere; because the masses need to be the same for the experiment.
d) Cube and disk; because the weights need to be the same for the experiment.
e) Sphere and disk; because the volumes need to be the same for the experiment.

a) Cylinder and block, because the materials need to be the same for the experiment.
b) Cube and block; because the volumes need to be the same for the experiment.
c) Block and sphere; because the masses need to be the same for the experiment.
d) **Cube and cylinder; because the weights need to be the same for the experiment**.
e) Sphere and cube; because the volumes need to be the same for the experiment.

The test item is related to the identification of independent and dependent variable (how weight or mass (volume in post-test) – independent variable influence the buoyant force – dependent variable) keeping the control variable constant (volume in pre-test, mass in post-test). In both pre and post-test many wrong answers were connected with answers that were containing the key words of mass or weight (in pre-test) or volume (in post-test). We believe that these words in bold students connected with the answer that students selected finally, even if it was wrong. This result suggests that students do not read the text with understanding. Many students also selected only the first correct answer, even though, if reading thoroughly, they would come across another correct one.

4. Conclusion

To sum up, we can conclude that the consistent implementation of inquiry activities across the three subjects of mathematics, physics and informatics has brought a significant improvement in the level of inquiry skills development. Focusing on activities in physics and testing the skills corresponding to physics, in particular, the impact in the subtest consisting of items having physics context was identified. Since the test items were closely connected with the skills dominantly developed in the physics lessons, we can conclude that the inquiry activities implemented in physics lessons influenced the learning gains in a positive way. However, we can identify some individual test items without any evident positive shift, even without conducting statistical comparison for individual test items. Surprisingly, students did not achieve any positive learning gains concerning the skills to formulate hypothesis and determining accuracy. The possible reasons of these results can be found in the lessons and the way how the instruction was conducted. Even though both skills were designed to be implemented within the activities, teachers, apparently, did not pay enough attention to them in the physics lessons. On the other hand, these skills were tested by a limited number of test items. In order to get more consistent results the further research will be aimed at detailed monitoring of the activities and the way how they are implemented in the physics classroom. The level of skills typically developed in the physics lessons will be diagnosed by the test designed for these particular skills covering each of them by several test items. Subsequently, the correspondence between the test results and inquiry activities with focus on particular skills development will be identified. We hope to develop the model of implementation of inquiry activities across physics lessons that will be proved to be appropriate and therefore applicable widely at upper secondary school level.

Acknowledgement
This work was supported by the Slovak Research and Development Agency under the contract no. APVV-0715-12 Research on the efficiency of innovative teaching methods in mathematics, physics and informatics education.

References

Banchi H., Bell R. (2008) The many levels of inquiry. *Science and Children* **46**(2).

Burns J.C., Okey J.R. & Wise K.C. (1985) Development of an integrated process skill test: TIPS II, *Journal of Research in Science Teaching* **22**(2), 169-177.

Harlen W. (2013) Assessment and Inquiry-Based Science Education: Issues in Policy and Practice. Trieste: TWAS-Strada Costiera.

Hattie J. (2009) Visible Learning. New York: Routledge.

Ješková Z., Lukáč S., Hančová M., Šnajder Ľ., Guniš J., Balogová B. & Kireš M. (2016) Efficacy of Inquiry-based learning in mathematics, physics and informatics in relation to the development of students' inquiry skills, *Journal of Baltic Science Education* **15**(5), 559-574.

Ješková Z., Kireš M. & Balogová B. (2018) Assessing inquiry skills of upper secondary school students. In *Proceedings of GIREP Seminar 2016* (Sokołowska D., ed.), Krakow, 30.August – 3.September 2016.

Linn M.C., Davis E.A. & Bell P. (2004) Internet environments for science education. Mahwah, NJ: Lawrence Erlbaum Associates.

Slovak national VEMIV project. Available on http://ufv.science.upjs.sk/_projekty/vemiv/projekt.php.

Trilling B. & Fadel Ch. (2009) 21st Century Skills: Learning for Life in Our Times. San Francisco, Jossey-Bass.

Wenning C.J. (2007) Assessing inquiry skills as a component of scientific literacy, *Journal of Physics Teacher Education Online* **4**(2), 21-24.

Wenning C.J. (2010) Levels of inquiry: Using inquiry spectrum learning sequences to teach science. *Journal of Physics Teacher Education Online* **5**(4), 11-19.

IOP Conf. Series: Journal of Physics: Conf. Series **1076** (2018) 012022 doi:10.1088/1742-6596/1076/1/012022

Assessing inquiry skills of upper secondary school students

Zuzana Ješková, Brigita Balogová, Marián Kireš

Pavol Jozef Šafárik Univeristy in Košice, Slovakia

Abstract. Inquiry-based science education (IBSE) has been in the focus of educational systems all over Europe in the last years. The main idea connected with IBSE is that students are involved in the process of learning by conducting activities to answer research questions similar way as it is done in science. This way they should gain not only scientific knowledge but also skills to do inquiry in order to understand how scientists work. In Slovakia this approach has been in force since 2008 when inquiry and the development of inquiry skills became the integral part of the science curriculum. With regard to the current science curriculum and increased interest in science and mathematics and informatics education a national project aimed at research on the efficiency of innovative teaching methods in mathematics, physics and informatics education has been running in the last few years in Slovakia. One of the project main research questions was to find out what is the current level of development of selected inquiry skills of upper secondary school students. In order to answer this research question a test assessing the level of selected inquiry skills was developed. The test was designed in cooperation with upper secondary school teachers. The test items are set in the context of physics as well as mathematics and informatics, however they test inquiry skills that should be content independent. The test was administered to 751 students aged 15-17 from the 1st and 2nd grade of upper secondary school. The test results have shown rather low level of students' achievements with the mean score of 32.5%. It has also revealed differences between classes of different age groups, different streams and gender. Surprisingly, the 2nd grade students did not achieve better results than the 1st grade students. In accordance with our expectations, classes with focus on mathematics and informatics achieved better results than general or language classes and boys were better than girls. There were also different achievements in the test items aimed at different inquiry skills with the best results achieved in the field of designing an experiment and identifying variables and weakest results achieved in the field of argumentation. In the paper the results of the test across the sample are discussed and analysed in detail, with focus on selected inquiry skills and the context of physics.

1. Introduction

Scientific inquiry is currently one of the most used terms in science education. There are many definitions of scientific inquiry. According to National Education Standards (NRC, 1996) „scientific inquiry refers to the diverse ways in which scientists study the natural world and propose explanations based on the evidence derived from their work. Inquiry also refers to the activities of students in which they develop knowledge and understanding of scientific ideas, as well as an understanding of how scientists study the natural world". Resulting from the definition, students conducting activities should gain not only scientific knowledge but also skills to do inquiry in order to understand how scientists work. This approach to science education (inquiry-based science education, IBSE) is in strong focus of European educational systems, involving Slovakia. It has been reflected in a number of projects at European and national level over the last years (e.g. 7FP: SAILS, ESTABLISH). In Slovakia the inquiry approach with focus on inquiry skills development is clearly defined in the curricular goals since 2008 not only in science but the elements of inquiry can be found in mathematics and informatics curriculum. With regard to the curriculum that has been in force already for several years, the following research question has emerged: What is the current level of development of selected inquiry skills of upper secondary school students? Accordingly, the purpose of the presented research is to diagnose the present state of development of selected inquiry skills without any external intervention to teaching in order to find out how the curricular goals have been reflected in the students' achievements. The presented research question represents the initial research question of the national project aimed at research on the efficacy of innovative teaching methods in mathematics, physics and informatics education (VEMIV). In the following paragraphs the answer to this research question is presented with regard to the methods of data collection and their analysis and results' discussion in the context of physics.

2. Methods

Development of a test of inquiry skills

In order to answer the research question about the current level of development of selected inquiry skills a test assessing selected inquiry skills was designed. Firstly, different frameworks of inquiry skills based on the stages of inquiry process or level of sophistication (Wenning 2005, 2010) with regard to age group (NRC, 2000) were analysed. Based on the existing taxonomies of inquiry skills reflecting the inquiry process (Fuhrman 1978, Tamir & Lunetta 1981, Fradd *et al.* 2001, Van den Berg 2013) we have adapted and agreed on the framework of inquiry skills with focus on experimental activities (table 1). This taxonomy served as a framework for the test design. We have also benefitted from the several available tests of inquiry skills, e.g. Scientific Inquiry Literacy Test, ScInqLiT (Wenning 2007), Test of Integrated Process Skills (TISP, Burns *et al.* 1985) a Test of Scientific Literacy Skills, TOSLS, (Gormally *et al.* 2012).

Table 1. Taxonomy of inquiry skills for experimental activities adapted on the basis of works of Fuhrman (1978), Tamir & Lunetta (1981), Fradd *et al.* (2001) and Van den Berg (2013). The skills selected for testing are in bold

Skill group	Skills
1. Conception, planning and design	1.1. Formulate a question, define a problem.
	1.2. Formulate hypothesis or expectation to be tested.
	1.3. Design experiment (which variables, which relationship).
	1.4. Design observation and/or measurement procedures (incl. lab-apparatus selection; experiment set-up) for each variable.
	1.5. Predict results of experiment.
2. Implementation	2.1. Manipulate apparatus/software.
	2.2. Observe/ measure.
	2.3. Record results.
	2.4. Calculate during the execution.
	2.5. Explains or makes decisions about experimental techniques.
3. Analysis and interpretation	**3.1. Transform results into standard form (i.e. tables, graphs).**
	3.2. Determine relationships between variables based on e.g. graphs, tables, text and formulas.
	3.3. Determine accuracy of experimental data (identify possible sources of errors).
	3.4. Compare experimental data to the hypothesis/ expectation.
	3.5. Discuss limitations/assumptions of the experiment.
	3.6. Propose generalizations of experiment results.
	3.7. Formulate new questions/ problems.
	3.8. Draw conclusion.

4. Communica-tion	4.1. Share and present results in front of the class.
	4.2. Discuss/ defend results/form arguments.
	4.3. Elaborate formal report about the gained results.
5. Application and follow-up	5.1. Predict on the basis of obtained results.
	5.2. Formulate hypothesis for follow-up.
	5.3. Apply experimental technique to a new problem.

The test was designed in cooperation with upper secondary school teachers. The individual test items are developed in order to measure the level of selected inquiry skills (table 1, in bold). The test items are set in the context of physics as well as mathematics and informatics. Firstly, a set of 12-16 test items was designed for each of these subjects. The test items were reviewed by physics, mathematics and informatics teachers who considered their difficulty and appropriateness of the test item to the corresponding inquiry skill that the test item is assigned to assess. Based on teachers' comments and revisions, the three reviewed subject tests were implemented in four classes of upper secondary schools. As a result, there was one test with test items selected from the subject tests developed. The draft version of the test was again implemented in a selected upper secondary class. After additional smaller corrections the final version of the test was prepared. The test includes 12 test items in order to be answered within one standard 45 minutes' lesson. Most test items are multiple-choice items with one or two correct answers and several items are open-ended. Examples of test items in the context of physics can be seen in tables 4, 5, 6.

Research sample

The test was implemented in the school year 2015/2016 in 11 upper secondary schools with 32 classes involving more than 800 students answering the test. From the whole research sample we gained 751 valid answers. The students included into the analysis belonged to 1st grade classes (337 students aged 15-16) and 2nd grade classes (414 students aged 16-17). The classes belonged to one of the three streams focusing either on foreign languages or the subjects of mathematics and informatics and the rest of the students belonged to classes with no special focus (general classes). The research sample involved 457 girls and 294 boys (table 2).

Data analysis

After collecting data from the test taken by the research sample, the analysis of test results was performed. The quantitative analysis involves basic descriptive statistics as well as statistical comparisons between different groups of students based on grade, gender and specialization. In order to compare achievements of different sample groups, the Shapiro-Wilk test of normality and subsequesntly, the Mann-Whitney test was applied. The statistical analysis was carried out in R software (http://www.R-project.org). In the qualitative analysis the students' answers were analysed with regard to the most common misconceptions and difficulties that students experienced when answering the test.

3. Results and discussion

Quantitative analysis

The mean test score for the whole sample reached the value of 32.5% (table 2). The mean score is much lower than expected, although the test items were not considered demanding based on teachers' evaluation of test items. This result indicates that not enough attention is paid to the activities that enable to develop these particular skills, even though it is clearly formulated in the curriculum.

Table 2. Basic statistics for the whole sample as well as for sample groups and comparison of different sample groups

	Whole sample	grade		gender		specialization		
		1st	2nd	boys	girls	Math/Inf classes	Lang. classes	Gener. classes
Number of students	751	337	414	294	457	95	330	326
Mean score (%)	32.5	31.4	33.4	36.3	30.0	42.1	30.2	31.9
Median (%)	31.3	29.2	33.3	31.3	33.8	41.7	29.6	31.3
St. dev. (%)	16.2	15.6	16.7	17.3	15.0	18.4	15.9	14.9
St. error of mean (%)	0.6	0.6	0.6	0.6	0.5	0.7	0.6	0.5
Test of normality (p-value)	< 0.001	< 0.001	< 0.001	0.003	< 0.001	0.61	< 0.001	
Mann-Whitney test		0.07		< 0.001		< 0.001		

Statistical analysis shows that neither the whole sample nor the sample groups are normally distributed ($p = 0.003$ for the group of girls and $p < 0.001$ for any other group), except from the math/inf classes ($p = 0.61$). As a result, the comparative statistics based on Mann-Whitney test was applied showing the following results. There is no significant difference between the 1st and 2nd grade students ($p = 0.07$). Even though second grade students have already experienced one more year of learning compared to first grade students, it seems that this extra year spent at upper secondary school had a negligible effect on their results that are similar to their novice schoolmates. This surprising result confirms that the first year upper secondary school education does not influence the level of development of these particular skills. On the other hand, there is a significant difference between boys and girls ($p < 0.001$) as well as between the math/inf classes and other classes ($p < 0.001$).

In table 3 the individual test items score are presented with regard to the corresponding inquiry skill as well as the context in which the item is put into. The items with physical context are marked bold. The last column shows the score for the particular skill group.

It can be seen that students achieved best results in the skill to design an experiment and identify variables (49.7%). It indicates that students conduct experiments, at least, according the prescribed procedure. However, all the other results are approximately somewhere between 28-46%, except for the skill concerning discussing and forming arguments were students gain the lowest score (7.6%). This particular test item requires choosing the correct answer and also forming arguments to justify the choice. This proved to be a demanding challenge for students who either did not give relevant arguments or skip the reasoning at all. This particular skill group 4 connected to defending results and forming arguments seems to be the most problematic for students. The best results were achieved in the skill group 1 connected to Conception, planning and design of an experiment (39.5%), while in the skill group 3 aimed at skills to interpret and analyse students achieved the mean score of 32.5%, and the mean score of 18.5% was achieved in the skill group 4. These results indicates that even though quite a lot of students are able to design an experiment and identify variables, they are much worse in data analysis and their interpretation with the worst results in the field of formulating coherent arguments.

Table 3. Mean test score achieved in individual test items corresponding to certain inquiry skill and skill group (Ph-physics, I-informatics, M-mathematics). The test items with physical context are in bold. The numbering of the skill groups corresponds to the taxonomy of inquiry skills (table 1)

Inquiry skill	Test item context	Mean score (in %)	Mean score (in %)
1. Conception, planning and design			
Formulate hypothesis or expectation to be tested	**Ph**	**30.7**	39.5
Design experiment (which variables, which relationship)	**Ph**	**49.7**	
Design experiment (which variables, which relationship)	**Ph**	**36.2**	
Design experiment (which variables, which relationship)	I	41.3	
3. Analysis and interpretation			
Transform results into standard form (i.e. tables, graphs)	I	28.6	32.5
Determine relationships between variables based on data in graph	**M/Ph**	**29.7**	
Determine relationships between variables based on data in table	I	29.7	
Determine relationships between variables based on data in table	I	45.8	
Determine relationships between variables based on data in text	M	31.7	
Determine accuracy of experimental data (identify possible sources of errors)	**Ph**	**29.6**	
4. Communication			
Discuss/defend results/form arguments	M	7.6	18.5
Discuss/defend results/form arguments	M	29.0	

Qualitative analysis
In the following section the specific test items are put into the context of physics are analyzed and discussed in detail with regard to most common revealed difficulties and misconceptions.

Inquiry skill *Formulate hypothesis to be tested*
In this test item (table 4, first part) students are expected to choose the correct hypothesis to test the question if the earth and earth are heated equally in the sunlight. Students achieved rather low score (30.7%). Many of them selected just one answer, mainly c. Common wrong answers were a and b. In the first case this is a question, not a statement, while the second choice is a hypothesis but not testing the question. This indicates that students do not have a proper understanding about what the hypothesis is.

Inquiry skill *Design experiment (which variable, which relationship)*
This test item (table 4, second part) with mean score of 49.7% is aimed at identifying independent and dependent variable to measure in the experiment aimed at heating different materials. The most common difficulties were related to confusion between dependent (temperature) and independent variable (time), selection of wrong variables, e.g. amount of sunlight, density, volume, mass, heat. Other problems connected with incomplete graphs, missing table or missing description of a variable were also quite frequent.

Table 4. Example of test item focusing on the skill to design an experiment (Burns *et al.* 1985)

Maria wondered if the earth and oceans are heated equally by sunlight. She decided to conduct an investigation. She filled a bucket with a kilogram of soil and another bucket with a kilogram of water. In sunny summer day she placed them so each bucket received the same amount of sunlight.
Which hypothesis should be tested by Maria to get the answer to her question? a) How is water and soil heated by the sun? b) The longer the soil and water are in the sun, the warmer they become. **c) Soil and water are heated by the sun differently.** d) Soil and water receive different amounts of sunlight at different times of the day. **e) Soil and water are heated equally.**
Maria is going to conduct an experiment using buckets with water and soil. Design a **table** and draw **the axes of a graph** with physical variables to be recorded by Mary to get the answer to her question. **Name the selected physical variables.**

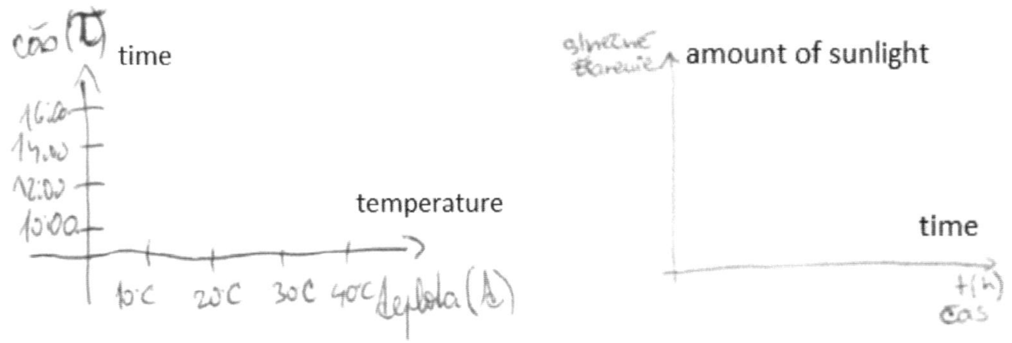

Fig. 1. Confusion between dependent and independent variable (left) and wrong selection of variables (right)

Inquiry skill *Determine relationship between variables based on graphs*

This test item (table 5) was aimed at graph interpretation reaching the mean score of approximately 30%. Most students were able to identify that the intersection of the two graphs represent the moment when one runner reaches the other one, however, to decide who reaches who was not so simple (distractor c). The wrong answer d was also quite common, i.e. students assigned higher average speed to the runner whose distance vs. time graph lies higher than that of the other. This belongs to a common misconception concerning confusion between variables of distance and speed in this case.

Inquiry skill *Determine accuracy of experimental data (identify possible sources of errors)*

This test item (table 6) with mean score of 29.6% was aimed at identifying possible sources of errors in data taken by videomeasurement, in this case. Nevertheless, the most commonly selected answer a) was not connected with reasoning the errors but it interpreted the presented graph, even in a wrong way. It seems that students, having a quick glance at the graph decided immediately about decelerated motion (declining graph) without deeper thinking and reading the text with understanding.

Table 5. Example of test item focusing on the skill to determine a relationship based on data in the graph

In the picture you can see a distance (from the start of the race) vs. time graph for two runners.

Decide about the correct statement:

a) The first kilometer was taken sooner by B.
b) A took more than 4 km in the first 10 minutes.
c) At a distance of 4 km from the start runner A reached runner B.
d) During the first 16 minutes A was running at higher average speed than B.
e) **B was running faster than A from the end of 10th min to the end of 11th min.**

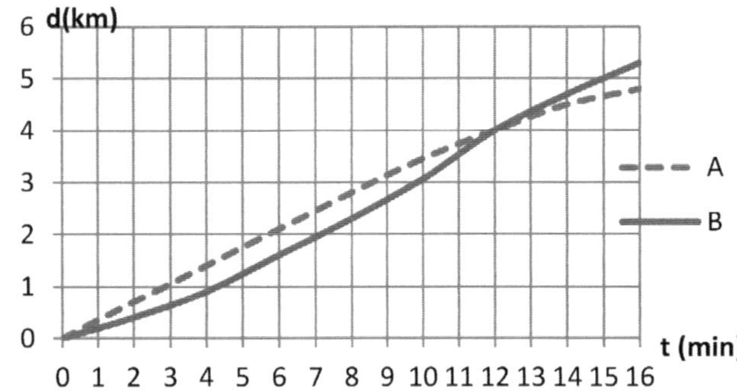

Table 6. Example of test item focusing on the skill to identify possible errors

In the graph the results of videomeasurement of a motion of a sailboat moving uniformly towards the lakeside is presented. Student has measured its distance from the lakeside clicking on the selected point on each videorecording frame while videoanalysing software record time and corresponding distance. He has found the best fit line. What is the **most likely reason** of the light scatter in the graphed data?

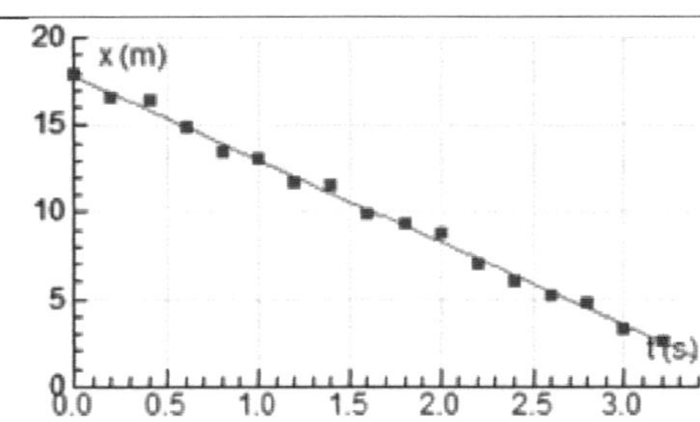

a) Sailboat does not move uniformly, it slows down.
b) **Student did not mark the selected point accurately on each frame.**
c) Video was not recorded precisely.
d) Sailboat does not move uniformly, it speeds up.
e) Software does not record time accurately.

4. Conclusion

To sum up the results of inquiry skills assessment it can be concluded that the level of development of inquiry skills is not satisfactory not reaching the average value of 40%. Surprisingly, no differences between older and younger students were identified, however, boys achieved better results than girls and the students of special mathematics/informatics classes achieved the best results. The best results were achieved in the skill group of designing experiment and identifying variables and relationships while there were problems identified concerning skills to analyse and interpret data. The worst results were identified in the skills of formulating hypothesis, identifying possible sources of errors and argumentation skills.

The level of development of inquiry skills of upper secondary school students diagnosed by assessment test indicate that students are likely not involved in inquiry activities that is an inevitable assumption of inquiry skills development. This is a signal for teachers to change their attitude towards the way how they teach and how their students learn in order to implement intentionally more student-centred and inquiry-based strategies.

Acknowledgment
This work was supported by the Slovak Research and Development Agency under the contract no. APVV-0715-12 Research on the efficiency of innovative teaching methods in mathematics, physics and informatics education.

References

Burns J.C., Okey J.R. & Wise K.C. (1985) Development of an integrated process skill test: TIPS II. *Journal of Research in Science Teaching* **22**(2), 169-177.

ESTABLISH project. Available on http://establish-fp7.eu/.

Fradd S.H., Lee O., Sutman F.X. & Saxton M.K. (2001) Promoting Science Literacy with English Language Learners Through Instructional Materials Development: A Case Study. *Billingual Research Journal* **25**(4), 417-439.

Fuhrman M. (1978) Development of a laboratory structure and task analysis inventory and an analysis of selected chemistry curricula, Unpublished master's thesis. University of Iowa.

Gormally C., Brickman P. & Lutz M. (2012) Developing a Test of Scientific Literacy Skills (TOSLS): Measuring Undergraduates' Evaluation of Scientific Information and Arguments. *CBE – Life Sciences Education* **11**(4), 364-377.

NRC, National Research Council (1996) National Science Education Standards. Washington, DC: National Academies Press.

NRC, National Research Council (2000) Inquiry and the National Science Education Standards, A Guide for Teaching and Learning, Washington, DC: National Academies Press.

SAILS project. Available on http://sails-project.eu.

Slovak national VEMIV project. Available on http://ufv.science.upjs.sk/_projekty/vemiv/projekt.php.

Tamir P. & Lunetta V.N. (1981) Inquiry-Related Tasks in High School Science Laboratory. *Science Education* **65**(5), 477-484.

Van den Berg E. (2013) The PCK of Laboratory Teaching: Turning Manipulation of Equipment into Manipulation of Ideas, *Scientia in educatione* 4(2), 74-92.

Wenning C.J. (2005) Levels of inquiry: Hierarchies of pedagogical practices and inquiry processes. *Journal of Physics Teacher Education Online* **5**(2), 3-16.

Wenning C.J. (2007) Assessing inquiry skills as a component of scientific literacy. *Journal of Physics Teacher Education Online* 4(2), 21-24.

Wenning C.J. (2010) Levels of Inquiry. Using inquiry spectrum learning sequences to teach science, *Journal of Physics Teacher Education Online* **5**(4), 11-19.

IOP Conf. Series: Journal of Physics: Conf. Series **1076** (2018) 012023 doi:10.1088/1742-6596/1076/1/012023

A proposed sequence of activities inspired by Galileo to exercise reasoning and modelling about motion and frames of reference

Marina Castells, Aikaterini Konstaninidou, Àngela Garcia-Lladó

Dep. Linguistic and Literary Education, and Didactics of Experimental Sciences and Mathematics Universitat de Barcelona, Catalonia, Spain

Abstrac. This paper presents a proposed teaching sequence by means of the description of teacher-mediated activities in a teaching sequence completed by primary and secondary education teacher trainees engaged in physics modelling. During the sequence, students solve a series of activities guided by the trainee teachers, whose discourse includes verbal and non-verbal communication and actions on or with physical objects, drawings, diagrams, graphs and other communicative resources. The students are guided through a process of knowledge building that begins with physical phenomena and advances through several levels of abstraction. This paper illustrates the process by analysing the narrative of the teaching sequence and the detail of some of the activities with the role of teachers and students in the physics modelling process.

1. Introduction: background and proposal

Currently, there is a general consensus that science education has to develop a broad range of competences in future citizens, including competences in science and technology. School-leavers must be able to use theoretical models to interpret scientific facts and think critically about issues of science and technology and about scientific information as it appears in the media. Moreover, as individuals they must share values related to science and technology that help them contribute directly to solving socio-scientific problems or negotiate technological issues with scientific and technological knowledge, methods and argumentation skills.

Related to this general context of education, conceptual learning acquires new importance. Students enter the classroom confident that the knowledge they already have can help them to interpret the world, but we know that part of this knowledge conflicts with scientific knowledge; and because some students are reluctant to reconsider what they think they know, teachers must convince them of the value of finding 'new ways of seeing' the world. According to Ogborn (1996), teachers have to create a need for knowledge in their students ("Creating Differences") that attracts them to new knowledge and helps them accept it. If the learning process involves many changes in the 'ways of seeing the world' (Leach & Scott 1995) teachers must be ready to guide students through these, using as their point of departure real phenomena or 'experimental facts' or activities presented in the classroom through dialogical discourse. And this discourse will be multimodal, making use of verbal communication, gestures, actions, drawings or diagrams and videos, among other resources, as well as arguments given in explanations or in argumentation with the students.

This paper considers the role of the teacher as a guide in the abstraction process that students must complete in order to build scientific models (Gutiérrez 2001). The aim of our study is to apply a proposed sequence in several groups of students and to observe through video recording how it works. In the implementation, the students are guided through a process of knowledge building that begins with physical phenomena and advances through several levels of abstraction. This paper illustrates the process by analysing the narrative of the teaching sequence and the detail of some activities and some emerging problems, with the role of teachers and students in the physics modelling process.

The teaching topic in all the problems will be Galilean relativity (Castells *et al.* 1996, Castells 1997, Saltiel & Malgrange 1980, Viennot 1996).

2. Theoretical framework

Our study is underpinned by three theoretical bases: the use of multimodal resources to make classroom explanations convincing (Ogborn *et al.* 1996, Kress *et al.* 2001); Perelman's rhetorical model of argumentation for justifying and convincing (Perelman & Olbrecht-Tyteca 1969, Perelman 1982); and the Science Modelling Cycle (Couso & Garrido 2016). These are described in detail below.

Explaining science in the classroom and multimodal resources (Ogborn *et al.* 1996, Kress *et al.* 2001). According to this perspective, science classroom explanations are communicative activities that have to be understood in their broadest sense as knowledge-building activities. Scientific meaning is built by means of language in its verbal, visual and formal mathematical modes, and with other semiotic instruments (e.g., gestural language and acts of communication based on or using objects from the physical world) that mediate the construction of new "explanatory entities" and models that help students think about and comprehend the real world or laboratory phenomena. The teacher uses various communicative resources to make their 'stories' convincing. (Màrquez *et al.* 2006, Clement 2008, Castells 2006).

The rhetorical and argumentative perspective (Perelman & Olbrechts-Tyteca 1969, Perelman 1982). This provides a very useful theory for understanding how teacher discourse uses rhetoric and argumentation. The theory consists of different discursive elements that contribute to the consistency of the discourse and its influence on a particular audience, or that provide the means by which to help convince this audience (Fagundez & Castells 2012, Castells *et al.* 2007).

Science Modelling Cycle (Couso & Garrido 2016). Science education experts have built models of learning cycles to describe science education within a constructivist paradigm comprising several steps in the students' construction of knowledge (Fig. 1) (Jorba & Sanmartí 1997). These steps go from simple to complex and from concrete to abstract along two axes. More recently, various cycles resembling the learning cycle have been adapted to suit the construction of scientific models or modelling processes in science classrooms (Gutierrez 2001). One of these is the new Science Modelling Cycle (Hernández *et al.* 2015, Couso & Garrido 2016), in which the steps in the students' learning cycle are differentiated from the teacher's instructional cycle in a classroom (Fig. 2).

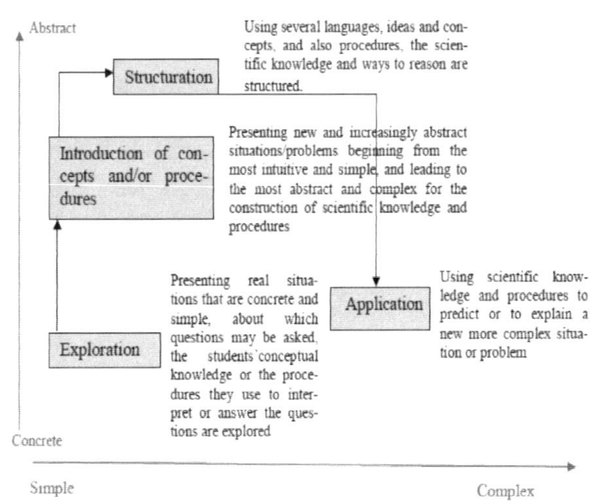

Fig. 1. Learning Cycle (Jorba & Sanmartí, 1996)

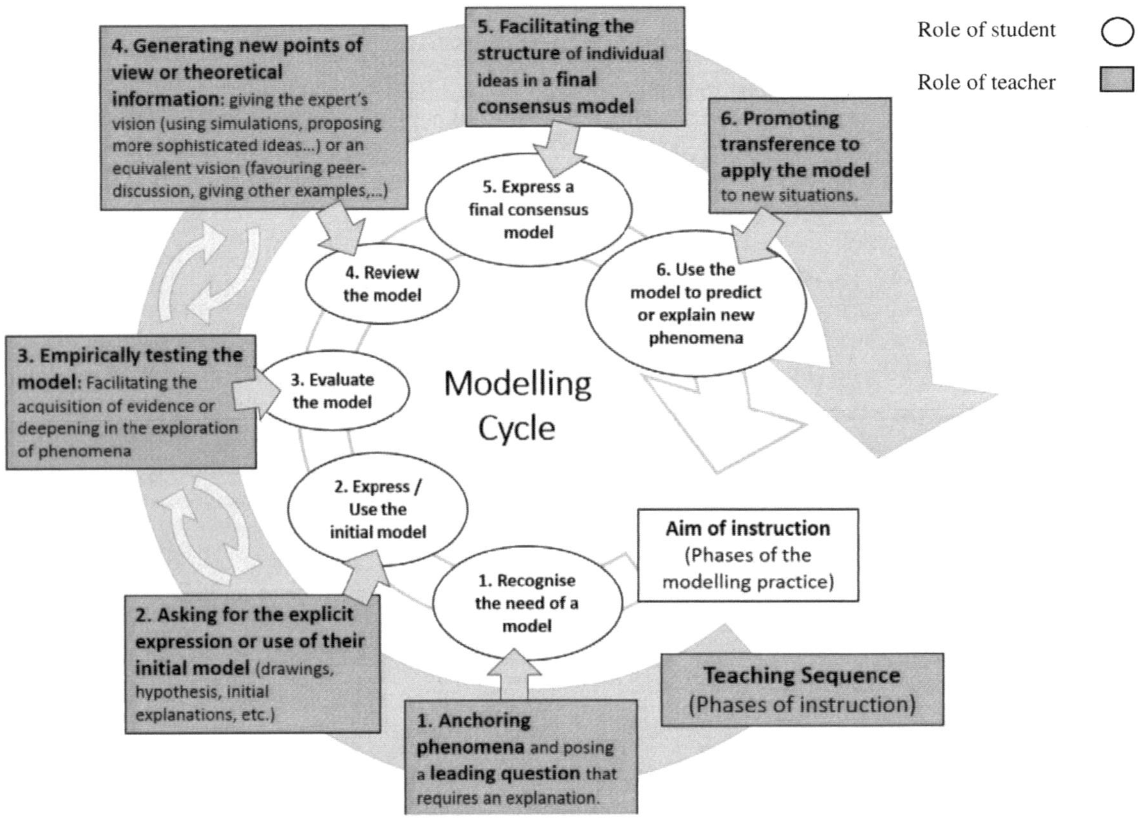

Fig. 2. Science Modelling Cycle (Couso & Garrido 2016)

How do these three theoretical bases contribute to our study? Ogborn *et al.* (1996) and Kress *et al.* (2001) understand that teachers' multimodal discourse in the science classroom is "built like a story" (Ogborn *et al.* 1996, p. 86), complete with protagonists and a plot, which helps us consider the steps of the Science Modelling Cycle as a sequence; and Perelman's rhetorical model of argumentation provides a very useful way to understand and evaluate the convincing character of the teachers' multimodal and argumentative discourse in the teaching-learning process. In the next page (Figure 3) we summarize the main aspects of the theoretical framework we used to complete the study.

3. Methodology

The focus of the research is a number of teacher-mediated activities in which the teacher uses a generally multimodal and dialogical teaching style to explain subject matter to a group-class (Scott *et al.* 2006). The teacher does this with hand-made or mental experiments, or demonstrations with actions on physical objects.

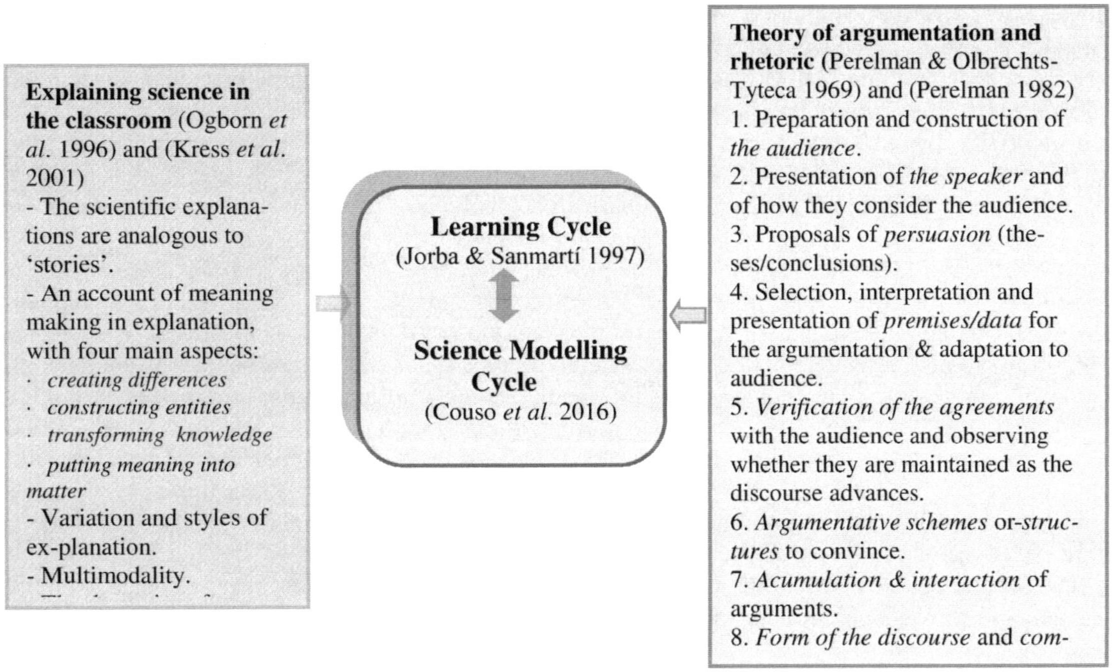

Fig. 3. Summary of our theoretical bases

However, we have organized the second part of this paper as a presentation of the narrative of the teaching sequence about Galilean relativity, in which some of the teaching are characterized in terms of the steps in the Science Modelling Cycle that they help the class to complete and also their degree of abstraction.

This methodology was chosen in line with our three theoretical bases, summarized above, which all contribute to our understanding of what occurs in the classroom in the teachers' instructional role. Bearing this in mind, the paper has actually been written as the case study of the activities (some are problems) in the teaching sequence rather than of the teachers themselves.

The classes take place in a physics laboratory and the experiments are generally performed by the teachers alone or with the support of the students. Our research is qualitative. The discourse of several trainee teachers is video-recorded during their multimodal intervention in physics classes.

The narrative of the teacher's explanation and the illustration of the activities in the teaching sequence

Once they have a teaching topic, the teacher or teaching sequence designer will need to choose a fundamental and inclusive question to support that topic's conceptual structure. This question, which we will call the 'big question', would be answered during the sequence or at its end.

In our particular sequence, the big question would be stated as follows: "Is motion absolute or relative to other bodies or systems?" Such a question would help the students understand the main principle of Galilean relativity, which is that 'the motion of a body (system) is always relative to another body (system) and absolute motion does not exist and has no scientific meaning'. To answer the question, however, the students would need to consider two further questions: "What is motion?" and "What is a frame of reference?"

Although the question "What is motion?" is not very difficult to answer, the scientific principle behind it is counter-intuitive, making it less easy for students to accept. The question "What is a frame of reference?" is more complex, chiefly because frame of reference (FR) is a highly abstract notion in physics. Consequently, the students' knowledge level will be important in determining which approach best introduces the notion of FR in the classroom. For example, for students in primary and compulsory secondary education the notion of Mechanical System (MS) introduced by Giordano Bruno (Tonnelat 1974) and Galileo may be more accessible than other approaches. In *La Cena de le Ceneri* (trans. *The Ash Wednesday Supper*) (1584), Bruno explains MS as "an ensemble of bodies animated with the same motion, but it is as if this motion in which they participate did not exist" and he also states that "it is impossible to recognize the motion of a Mechanical System from experiments made inside it" (Bruno, *La Cena de le Ceneri*, third dialogue). This also appears in Galileo's *Dialogues*.

In the case of trainee teachers in both the primary and upper secondary school classroom, the notion of FR put forwards by Bruno and Galileo could be expanded upon with Einstein's treatment of FR, mainly based on certain texts that clearly illustrate the physicist's notion of the concept (Jammer, with foreword of Einstein, 1970). A frame of reference is a space linked to an object of reference that we have to think of as being composed of infinite points radiating outwards from the object in all directions and that is without limits; the space moves with the motion of the object of reference and offers no resistance to the motion of any body. To describe motion it is also necessary to consider time, and we can imagine this space with a clock at every point. The notion of FR put forwards by Einstein is more abstract that Galileo's but helps students understand certain aspects of the relativity of motion more effectively.

The section below describes the outline of the conceptual plot of the teaching sequence for trainee teachers and for students in secondary education, which is inspired by the history of relativity (Tonnelat 1974).

- Motion is the change in position of an object in relation to someone or something.
- Objects that do not change position in relation to other objects constitute a reference frame. (Equivalently, we can say that all objects that move in the same way as others constitute a reference frame). The motion of any body is the same for all these objects.
- The description of any motion is always relative to a reference frame.
- The motion of an object can be described in relation to several reference frames. Motion changes when the FR changes.
- All objects that constitute an FR move in the same way in relation to another FR (in particular, the carrier and the carried object). Their mutual relations are such that this shared motion appears not to exist. (This appears in Galileo's *Dialogues*).
- The common motion of all the objects that constitute an FR is not lost when one of the objects is separated physically from the rest. (The motion that one object has by participating in the carrying motion is not lost when it becomes separated from the carrier.)
- If FR A moves in relation to FR B, FR B moves in relation to A with the opposite motion. This means that, in relation to each other, the motions are the same but opposite.
- We can interpret the motion of an object in relation to FR A as its motion in relation to FR B combined with the motion of FR B in relation to A.

Fragments from Galileo's *Dialogues* and Bruno's *La cena de le ceneri* would be adapted to provide a context or the inspiration for the activities (problems) proposed in the teaching sequence.

The activities: relationship with the Science Modelling Cycle and their abstraction level

At first, the students are presented with some qualitative activities (problems) designed to explore the students' conceptions about Galilean relativity, the first of which is the problem of 'the man in the river', which poses two questions (activities 1 and 2).

Activity 1. *The man in the river.*

> Imagine you are drifting down a river on a tyre. Ahead of you, a few metres downstream, there is a wooden box, floating on the water like you; and behind you, a few metres upstream, there is another box.
> Question 1. If you go on drifting, without swimming, will the distance between you and each box change with time? Why or why not?
> Question 2. Suppose the tyre gets flatter. Which of the two boxes will you reach first? Why?

The students solve this problem, first individually and in writing, and then together with their classmates, focusing on the 'Why?' and 'Why not?' questions and attempting to justify their answers. The teacher discusses the problem with the students, focusing on Question 1 to create the need for an initial scientific model.

Activity 1 would help the class complete the first step in the Science Modelling Cycle (teacher's role: anchoring phenomena, meaning the initial situation or problem; students' role: recognising the need for a model). Activity 1 also corresponds to what we could call a zero level of abstraction.

Activity 2. *Dialogue about students answers to Question 2 of the initial model in Activity 1*

Two student answers to Question 2 of Activity 1 are recorded below. Writing the answers to this question in groups is a way of completing the second step in the Science Modelling Cycle (teacher's role: asking the students for the explicit expression or use of their initial model) and also the first level of abstraction in the learning cycle, insofar as the students need to use words in a coherent discourse.

S1: *Even if the box upstream were nearer to us, swimming against the current requires a great effort and it would be a waste of time, so even if the box downstream were further away, we* ***would reach it first, because the speed we acquire by swimming is complemented by the impulse of the current****.*

S2: I *would go towards the box upstream because although I have to swim against the current,* ***the same current would bring the box towards me****; in contrast,* ***if I swam towards the box downstream I would take longer to reach it because at the same time as I swam towards it, the box would be moving downstream*** *by the force of the current.*

Students can offer various solutions but in many cases their arguments will only be based on their intuition, meaning in this case a (non-scientific) model they initially brought into the classroom.

Activity 3. *Motions from different FRs using slides as a visual resource*

This is related to activities 1 and 2. Teachers may help the students to build the "correct" solution using slides as a visual resource (Fig. 4).

With the help of a dialogical discussion about some of their answers to the initial problem and using the slides resource with Activity 1, students express and use their initial model about motion and FR, which helps the class complete the second step in the Science Modelling Cycle (teacher's role: predict the expression of the initial model; students' role: express and use the initial model) and the first level of abstraction in the Learning Cycle (the use of words only).

The slides resource consists in drawing one or several bodies or a place on each slide, representing possible FRs, and then moving the slides in relation to each other, so that the different motions can be "seen" in relation to one another. This is an effective way to construct the concept of FR based on Bruno's Mechanical System (Bruno 1584). (Note that slides offer a very adaptable resource in a variety of problem-solving activities related to Galilean relativity.)

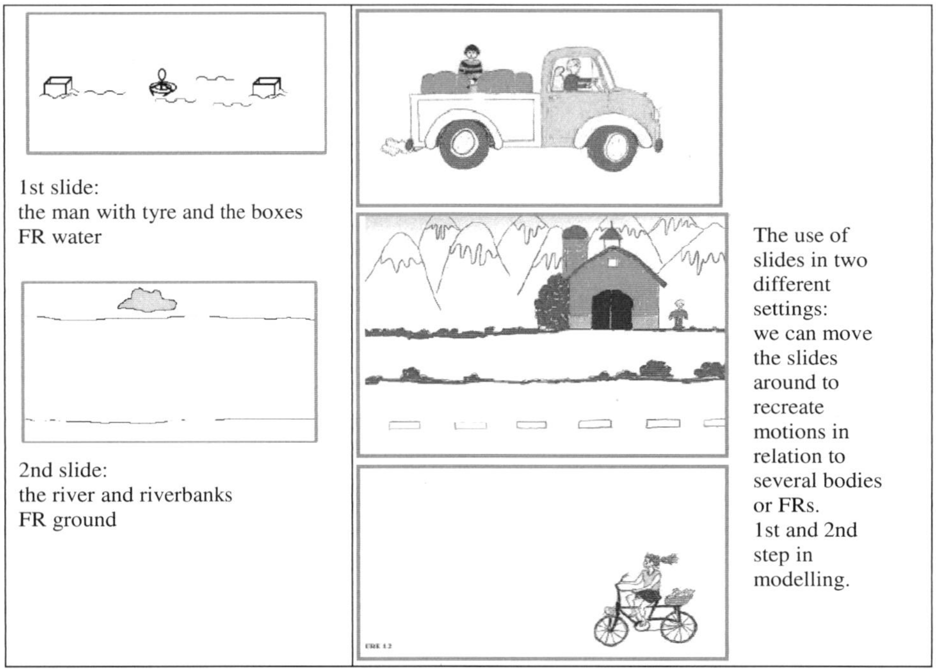

Fig. 4. The transparency recourse in the study of the relativity of motion

Activity 4. *The ball and the horseman*

This is inspired by Galileo's *The ball that runs behind the horseman* (*Dialogues* 1632, p. 156), which it adapts as a problem given to the students in written form (next page).

The activity helps the class complete the third step in the Science Modelling Cycle (teacher's role: empirical testing of the model; students' role: evaluating the initial model) or the fourth (teacher's role: generating new points of view or theoretical information; students' role: reviewing the model), as well as the first or second levels of abstraction in the Learning Cycle, respectively. The activity could be tailored to suit one particular step in the Science Modelling Cycle or in the abstraction process in one group class, and another step in another class. In fact, the steps are not characteristics of the activities themselves but of the specific use of the activity in a group class.

The activity could also be used to help the class complete the fifth step in the Science Modelling Cycle (teacher's role: facilitating the consensus model; students' role: expressing a final consensus model) or the sixth (teacher's role: promoting transference to apply the model; students' role: using the model to predict or explain new phenomena).

A copy of the text from Galileo's *Dialogues* could be given to students before or after they see the written problem.

One of the passages from Galileo's Dialogues about the two Great Systems of the World (1632) is about a ball that "runs behind" two horsemen.

a) In this passage, the first time the two characters disagree is about what would happen to a ball dropped by a horseman. Salviati argues that when the ball hit the ground, instead of stopping there it would go on moving forwards like the horseman; and that if it wasn't for the friction between the ground and the ball, the ball would go on moving forwards at the same speed as the

IOP Conf. Series: Journal of Physics: Conf. Series **1076** (2018) 012023 doi:10.1088/1742-6596/1076/1/012023

horse. Simplicius argues that the ball would stop still and end up behind the horseman, unless he had thrown the ball hard in the direction he was going.
Who is right? Justify your answer using the model of Galilean relativity we made in class.

b) The second time they disagree is when Salviati proposes that if one of the horsemen turned on his horse, while riding at a constant speed, and threw a ball backwards, in the direction he had come from, when that ball hit the ground it would continue to travel forwards or else simply stop where it fell, but would only continue moving in the opposite direction to the horseman if he had thrown faster than the horse was galloping.
Is Salviati right? Justify your answer using the model of Galilean relativity we made in class.

The idea could be adapted to make the task easier, as shown in Activity 5 below.

Activity 5. *The ball and the boy*

The students are given the same problem as in Activity 4 but Galileo's horseman is replaced by the figure of one of their classmates. The result is that it becomes easier to see how the ball follows the boy.

A student walking across the classroom in a straight line drops an iron ball. What happens to the ball when it hits the ground? Does it move? If it does, how?

The problem is based on Galileo's text and experimental activities 5 and 6 may help the students accept the idea that the motion gained by one object by participating in the motion of the carrier is not lost when this object is separated from that carrier (first concept: motion is not lost when an object is separated from its carrier). This is a very important idea in the new model. Here, the resource is a simple experience and the use of analogous cases.

If we put the focus on the first concept, Activity 5 could help the class complete the first step in the Science Modelling Cycle (teacher's role: anchoring phenomena, meaning the initial situation or problem; students' role: recognising the need for a model) and would also correspond to the zero level of abstraction. But it could also help them in the sixth step (teacher's role: promoting transference to apply the model; students' role: using the model to predict or explain new phenomena) (analogous to the activity using Galileo's text).

Activity 6. *The ball and the boy video-recorded*

This activity is like Activity 5 using the video-recording resource (Fig. 5) from two different FRs.

If the focus is on the first concept, Activity 6 could be used to help the class complete the second step in the Science Modelling Cycle (teacher's role: predict the expression of the initial model; students' role: express and use the initial model). But it could also be used for the sixth step (teacher's role: promoting transference to apply the model; students' role: using the model to predict or explain new phenomena) (analogous to Activity 4) and the first level of abstraction. The photograph of the trajectory seen from two FRs is like a representation of the trajectories of the ball seen from two different FRs.

IOP Conf. Series: Journal of Physics: Conf. Series **1076** (2018) 012023 doi:10.1088/1742-6596/1076/1/012023

Situation: A boy walks across the laboratory and drops a ball.

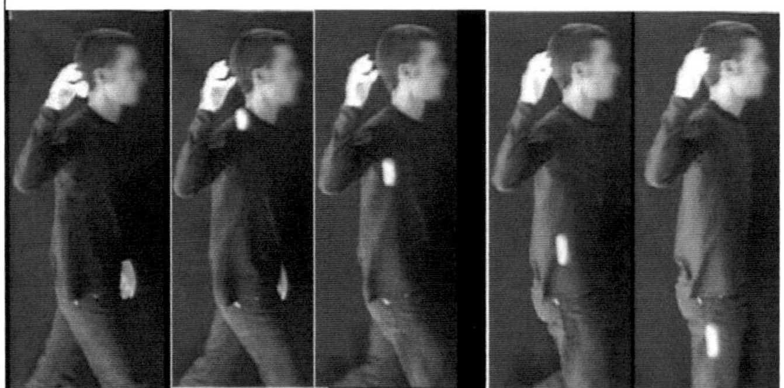

a) Video of a ball falling seen from the FR boy

The same situation: A boy walks across the laboratory and drops a ball behind a screen.

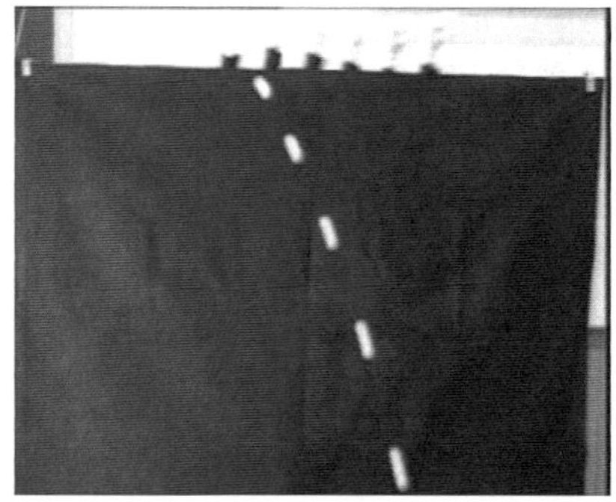

b) Video of a ball falling seen from the FR laboratory

Fig. 5. Activity 6 with the video-recording of a ball being dropped by a boy walking across the laboratory, seen from two different FRs

Activity 7. Representing the motion from two FRs using event diagrams

Fig. 6. The scenario of *The man in the river* (initial situation in relation to the bank)

Fig. 6 offers one way of illustrating the problem in activities 1 and 2, *The man in the river*. Fig.7 and 8 (bank and water, respectively) are two event diagrams to be used to answer Question 2 in that problem and to help the student appreciate the two FRs.

The event diagram resource may be tailored to support study at a variety of specific or abstract levels, depending of the students' age and class year, using realistic drawings, stick drawings or graphemes and symbols like dots and arrows.

Below (Figures 7 and 8), we offer an event diagrams for Question 2 with the FR riverbank or ground, and the FR water.

The event-diagram in a more abstract way is not included here, but it may be used specially with older students.

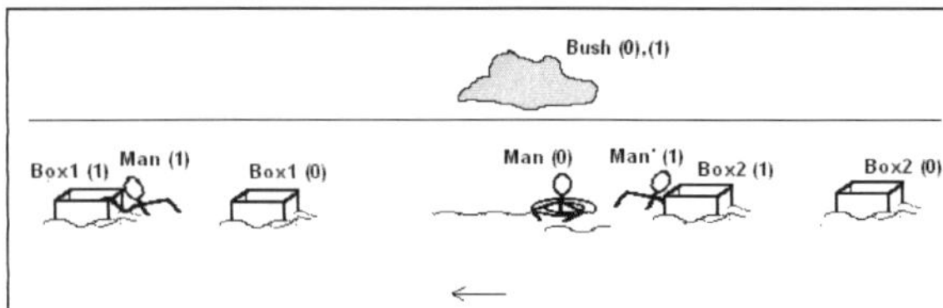

Fig. 7. Event diagram of Question 2 of the problem from the FR ground

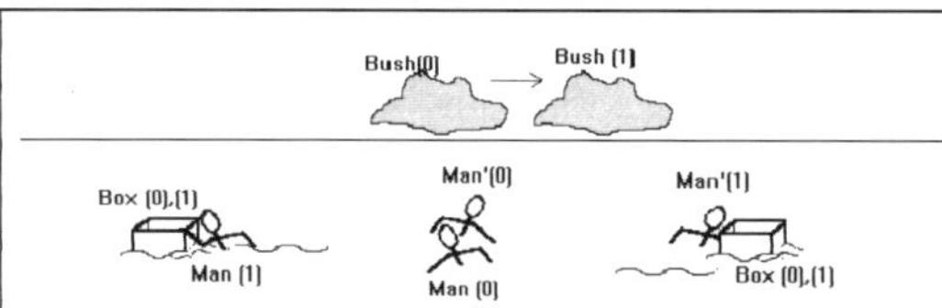

Fig. 8. Event diagram of Question 2 of the problem from the FR water (river)

Activity 7 helps the class complete the fourth step (teacher's role: generating new points of view or theoretical information; students' role: reviewing the model) or the fifth (teacher's role: facilitating the structure of individual ideas in a final consensus model; students' role: expressing a final consensus model), as well as the second level of abstraction. The diagrams *per se* are abstract representations that help the students construct the model of Galilean relativity. This resource is inexpensive and easy to use, and very useful for helping students "see" motion from different FRs.

Other activities that use new resources

Activity 8. *Tintin and relative speed*

This activity 8 uses the resource of a comic strip and may correspond to the sixth step of the use of the model constructed to predict or explain new phenomena (Figure 9).

Fig. 9. Activity 8. *Tintin and relative speed.* The comic as a resource to model relative speed

Activity 9. Recording trajectories with sheets of white paper and carbon paper

This activity can help the class complete the fourth, fifth or sixth steps, depending how it is used by the teacher. The new model is completed by adding the trajectories collected as convincing "proof" of the idea of the relativity of trajectories to the FRs, which is part of the scientific model being constructed.

Like Galileo, the teacher use analogies, asking the students to think of a rolling ball on the bed of a railway freight car, conveyor belt or other such surface characterized by uniform motion in relation to the ground. We propose an experimental arrangement, analogous to the situation described, using as resources sheets of white paper and carbon paper mounted on wooden boards and representing the FR train and the FR ground.

This activity reinforces the concept of the relativity of the trajectories to the FR using the resource of collecting trajectories (here, using moveable wooden boards and sheets of white paper and carbon paper). The activity helps the class complete the fifth and sixth steps in the Science Modelling Cycle. The trajectories and the arrows on them contribute to the third level of abstraction in the Learning Cycle.

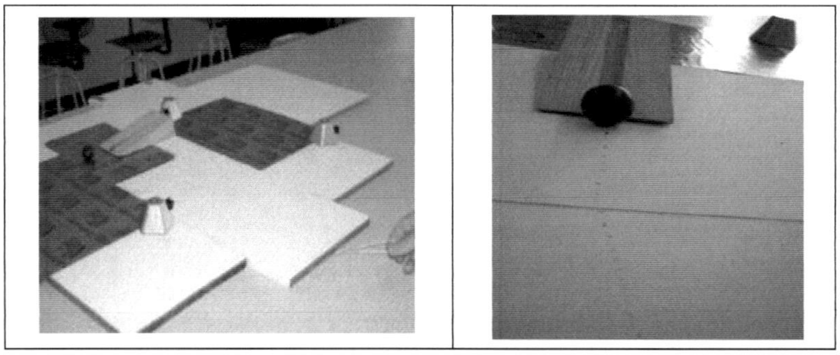

Fig. 10. Recording trajectories using sheets of white paper and carbon paper

4. Final discussion

The activities proposed and inspired mainly by the history of science, specifically by the *Dialogues* of Galileo and implemented by several teachers support the learning about the main ideas and concepts of the scientific model of motion. The activities may be used at a lower or higher level of abstraction depending on the aims of the educational levels (secondary education or teacher trainees in primary or secondary science education). The problem we need to address is how teacher education courses should be redesigned if we consider the didactical meaning of the Science Modelling Cycle or the Learning Cycle, as well as the need to include the abstraction process in the construction of the scientific model.

Only some of the many possible activities are proposed here and our aim has been to give some idea of the approaches that can be adopted and of the simple and inexpensive resources that can help pre-service science education teachers support the construction of scientific models in the classroom.

The study also shows that the subject history of science can provide a highly appropriate context for inspiring instructional activities that favour science knowledge transfer.

Acknowledgements

The research was supported by ARCE 2015-UB to GRIEC-UB, and by the Spanish EDU2013-47599-C2-2-P (Coord. Mercè Garcia-Milà) and the Catalan PRI 2014SGR1492 (LICEC 2014–UAB) (Coord. Mariona Espinet).

References

Bruno G. (1584) La Cena de le Ceneri. English translation (1966): *The dinner of the Ash Wednesday.*

Castells M. (1997) *Patrons de comportament dels estudiants en resoldre problemes de relativitat galileana, i factors que els influeixen en les respostes i en els raonaments.* PhD., Servei de Publicacions, U.A.B., Bellaterra, Barcelona.

Castells M. (2006) ¿Qué podemos aprender sobre las explicaciones de los profesores partiendo de una perspectiva retórico-argummentativa – comunicativa? In Atas. *Encontro Nacional de Pesquisa em Educaçao em Ciências* **5**(1-16), Baurú: Brasil.

Castells M., Cabellos M. & Cerveró J.M. (1996) Els *Diàlegs sobre els dos grans Sistemes del Món* de Galileo Galilei; un exemple vàlid en el context actual en l'ensenyament de la física. In IV Trobades d'Història de la Ciència, Societat Catalana d'Història de la Ciència i de la Tècnica. Barcelona: Societat Catalana d'Història de la Ciència i de la Tècnica (IEC), Alcoi, 487-498.

Castells M., Enciso J., Cerveró J.M., López P. & Cabellos M. (2007) What can we learn from a study of argumentation in the students' answers and group discussion to open physics' problems? In *Contributions from Science Education Research* (Pintó R. & Couso D., eds.), Dordrecht: Springer, 417-431.

Clement J.J. (2008) Depictive Gestures and Other Case Study. Evidence for Use of Imagery by Experts and Students. In *Creative Model Construction in Scientists and Students. The Role of Imagery, Analogy, and Mental Simulations*. Springer Science + Bussiness Media B.V., cap. 12, 171-204.

Couso D. & Garrido A. (2016) Models and modelling in elementary school pre-service teacher education: why we need both. In *Cognitive and Affective Aspects in Science Education Research* (Hahl K., Juuti K., Lampiselkä J., Uitto A. & Lavonen J., eds.) 11th ESERA Conference Selected Contributions. Springer Editors, 263-278.

Fagúndez Th. & Castells M. (2012) La argumentación en clases universitarias de física: una perspectiva retórica. *Enseñanza de las Ciencias. Revista de investigación y experiencias didácticas* **30**(2), 153-174.

Galileo Galilei (1632) *Dialogo di Galileo Galilei Linceo sopra i due Massimi Sistemi del Mondo Tolemaico e Copernicano*, English translation (1966) Dialogue Concerning the two Chief World Systems, with foreword of Einstein, A., USA.

Gutierrez R. (2001) Mental Models and the fine structure of conceptual change. In *Physics Teacher Education Beyond 2000* (Pintó & Suriñach, eds.), Paris: Elsevier, 35-44.

Hernández M.I., Couso D., Pintó R. (2015) Analyzing students' learning progressions throughout a teaching sequence on Acoustic Properties of Materials with a model-based inquiry approach. *Journal of Science Education and Technology* **24**(2), 356-377.

Jammer M. (1970) *Concepts of Space*, Foreword by Albert Einstein, Harvard University Press, Cambridge, Massachusetts, Second Edition.

Jorba J. & Sanmartí N. (1996) *Enseñar, aprender y evaluar. Un proceso de regulación continua.* MEC. Madrid.

Kress G.R., Jewitt C., Ogborn J. & Tsatsarelis C. (2001) *Multimodal science teaching and learning: The rhetoric of science classroom.* London/New York: Continuum.

Leach J. & Scott P. (1995) The demands of learning science concepts: Issues of theory and practice. *School Science Review* **76**(277), 47-51.

Márquez C., Izquierdo M. & Espinet M. (2006) Multimodal science teacher's discourse in modeling the water cycle. *Science Education* **90**, 202-226.

Mattews M.R. (1991) Science Teaching. The role of History and Philosophy of Science, London and New York: Routledge.

Ogborn J., Kress G., Martins I. & McGillicuddy K. (1996) *Explaining Science in the Classroom.* Buckingham, Open University Press.

Perelman Ch. (1982) *The Realm of Rhetoric*. Notre Dame (Indiana): University of Notre Dame Press.

Perelman Ch. & Olbrecht-Tytecha L. (1958) *La nouvelle rhétorique: traité de l'argumentation.* Bruxelles: l'Université de Bruxelles. English translation (1969) *The new rhetoric. A treatise on argumentation.* Notre Dame (USA): University of Notre Dame Press.

Saltiel E. & Malgrange J.L. (1980) Spontaneous' ways of reasoning in élémentary Kinematics, *European Journal of Physics* **1**, 73-80.

Scott Ph., Mortimer E.F., Aguiar O.G. (2006) The tension Between Authoritative Dialogic Discourse: A Fundamental Characteristic of Meaning Making Interactions in High School Science Lessons. *Science Education* **90**, 605-631.

Tonnelat M.A. (1974) *Histoire du Principe du Relativity*. Paris: Flamarion.

Viennot (1996) *Raisonner in Physics. Raisonner in Physics: La part du sens common*, Practiques Pedagogiques, Paris: De Boeck and Larcier.

IOP Conf. Series: Journal of Physics: Conf. Series **1076** (2018) 012024 doi:10.1088/1742-6596/1076/1/012024

Informal method of the research skills improving on the example of students' projects

Nataliya Kazachkova[1], Volodymyr Kurnosov[2], Julia Kurnosova[2], Vitalii Yurko[3], Uliana Nyemchenko[4]

[1]Scientific Physics and Technology Centre (SPTC) Ministry of Education and Science in Ukraine in Karazin Kharkiv National University
[2]B. Verkin Institute for Low Temperature Physics and Engineering NASU, Kharkiv Ukraine
[3]Kharkiv Gymnasium No 45
[4]Karazin Kharkiv National University

Abstract. It is well-known that appreciation of physics beauty starts since the early school age (usually from 7-12). At ECYGDA laboratory (which is an NGO situated at Karazin Kharkiv National University) STEM teachers together with students have been working on a vocational physics course for primary and secondary school children since 2007. The developing of research skills has been demonstrated on the example of two projects: Heron Fountain model and The Droplets Hovering Above The Vibrating Liquid which has been done by primary and secondary school students from the Centre.
The main goal of the activities described below is to trigger a kids' motivation to do simple physics research projects beyond the school curriculum. During such kind of regular laboratory trainings pupils have the opportunity to obtain an insight into scientific methods of investigation, to conduct their own projects, and promoting their activities and investigations at the different local and international conferences.

1. Introduction

At ordinary Ukrainian secondary schools physics lab (both in lyceums and gymnasiums) experiments and demonstrations during the lessons are carried out with pre-assembled equipment. More than 60% of Ukrainian schools have got serious lack of equipment at the lessons (Kazachkova *et al.* 2000, Kazachkova 2007); Moreover, in most cases the experiments are demonstrated using pre-assembled instructions. In spite of the fact that those activities are supported by Ukrainian National Doctrine of Science Education, students who are gifted in experimental approach should also have the opportunity to choose and carry out the projects they have thought up themselves. Similar problems have been mentioned in some articles such as (Trna 2005, Premier 2006, Dvorak 2007). For that reason the ECYGDA Laboratory has been created. That is a centre for the primary and secondary school students of three age groups (7-11, 12-15 and 15-18 years). The centre offers a special support for realising simple research projects. The SPTC staff encourages children to take part in annual local University and international Conferences to report about their achievements. Since ECYGDA has not got a financial support from the Government however there are three main sources of an income: parents' monthly beneficent payments (30% of total expenses), support from some Kharkiv Industrial Factories (40%) and private firms (30%).

IOP Conf. Series: Journal of Physics: Conf. Series **1076** (2018) 012024 doi:10.1088/1742-6596/1076/1/012024

There are some reasons for the faculty of Physics and Technology to participate at ECYGDA activities: to find out the pupils and secondary school students who are gifted for the STEM subjects, select and bring up them as future university students and prepare young generation to the future research work.

One of the main problem of Physics education in Ukraine since 1998 that the number of lessons a week was being decreased. For example since 2007 secondary school students have got 1 hour of physics a week instead of 2 even at the beginning of learning (6-7 forms). The situation for the senior classes are similar: from 4 to 2 (in 8 and 9 forms) and from 5 to 3 (at 10 and11 forms). However the demands of the official school curriculum to the students' skills and knowledge hold the same so there is a contradiction between a possibility for teachers to give their students the necessary knowledge and experimental skills under the stipulation of lack of time for that. From the other hands students who keen on physics and their parents wanted to stimulate their interest so creation of the ECYGDA laboratory had been actual and relevant.

2. Methodology

How the extracurricular trainings have been organised. The ECYGDA has agreements with 25 Kharkiv Secondary schools, where science communicators from the Centre with the help of University lecturers and students from the Faculty of Physics and Technology regularly demonstrate Physics Theme Shows, in a format of entertaining physics competitions related to the content of the Official School Physics Curriculum. During those shows the lecturers are able to select and choose the students who have capabilities for experimental work and invite them to join regular trainings on Saturdays at the Laboratory. Those selected primary and secondary school students have regular (once a week) short theoretical lectures (45 or 60 min), giving by university teachers accompanied by practical training (90 min) led by university teachers or students. In addition all our students have special English course (two hours a week), where they learn Physics and Maths in English. It is a very important point of their preparation as future scientists. It is considered there are three stages of experimental skills development.

The first stage is for all invited primary school pupils from 7-11. There are usually two groups of 20-25 pupils at that stage. At our theoretical training we proposed them 13 interactive theme physics lectures which have been elaborated by the teachers from the Centre. All of them have been adapted to the primary school pupils to be understandable for children of that age range. Every Saturday at the premises of the Centre one of the lectures (dur. 45 min) is presented to our visitors. The topics are interesting for children: Physics in Toys, Wonderful Mechanics, Travelling in Sound Land, Physics in the Kitchen, Light and Colours, Paradoxes of Magnetic Fields, Wonders of Electricity etc. At the beginning visitors became acquainted with simple physics principles and laws and then they are able to do simple experiments themselves. After 5 months training they choose a topic and prepare their own simple research projects. They usually report about their first "scientific results" at the annual University Conference "Junior Scientific Start-Up" in May. At the first stage they usually do simple experiments which are demonstrated and explained to the audience at the Conferences. In 2016 the best project was: Light and Sound waves with the help of simple models and toys (prepared by two primary school students).

The second stage is for students from 12 to 15 who are selected by methods mentioned above from Kharkiv schools and lyceums. Every year we have one group of 30 persons at theoretical training and not all of them have managed to finish their project to the end. They are also involved in regular extracurricular (once a week on Saturdays) short theoretical lectures (45 min) and more serious practical training (90 min). During such experimental training students are taught to use simple tools like handsaw, boring mill, perforator, vernier callipers, testers. They design and produce some exhibits for Physics Exhibition (Kazachkova 2007) or for the events which are organised in their schools (Week of Physics, Science Picnic, Night of Science) under the leadership of university students from the Department of Physics and Technology and research engineers from the Scientific Physics and Technology Centre. They gain a lot from such practical trainings and their experimental skills are seriously improved by doing self-made experimental projects using recycled materials, simple household objects or ordinary toys. In Fig. 1 you can see and compare the self-made exhibit called Heron Fountain which had been created at the first stage with the help of ordinary plastic bottles and the "Fountain" designed and produced at the second stage by the same student two years later. Those pictures give the opportunity to demonstrate and estimate how the student's research skills have been developed during two years of vocational trainings mentioned above.

Fig. 1. The photo (left) demonstrates the project made at the age of 10 after one year training at the Laboratory, second (in the middle) demonstrate the learning to use tools and the third one – the result of the work done at the age of 13 by the same student

The third stage of theoretical and research skills development. The prevailing lack of interest in physics matters among adolescents from 14 to 18 is obvious and common not only for Ukraine but also for all developed countries (Sjoberg & Schreiner 2006, Trna 2005). It most notably manifests itself in the steady decline in the number of students at Physics Departments at all Ukrainian universities. So the number of students who stayed to do research projects goes down to 10-15. ECYGDA with its location at the University mentioned above combined with the possibilities, associated to this fact – use of the machine laboratories and the electronic repair laboratories at the SPTC, subject-specific support by scientists, lease of equipment has got lots of advantages not only in Kharkiv Region, but also in Ukraine. The best research project of this year was done by a 15 year- old student who had not got an additional theoretical maths preparations. But his experimental skills had been developed through four years of trainings and it was his second research project. It was chosen by the student himself from the list of problems proposed by the International Young Physics Tournament, IYPT. From our point of view the problem can be useful and interesting to repeat during vocation training in any country so the detailed description will be proposed below.

3. The droplets hovering above the vibrating liquid

Problem formulation. "Hydrophobic Water" sets a dish filled with soapy water onto a loudspeaker or other vibrator. When it oscillates, it is possible to hold small droplets on its surface for a long time. Explain and investigate the phenomenon.

Literature analysis. Hydrophobic liquids are those ones, which surfaces are extremely difficult to wet. Droplets of such liquids have to float above the free liquid surface. Paradoxically, if a droplet falls down on the surface of vibrating water it can fully rebound like an elastic ball and then the droplet floats above the surface for a long time just like a hydrophobic one.

IOP Conf. Series: Journal of Physics: Conf. Series **1076** (2018) 012024 doi:10.1088/1742-6596/1076/1/012024

The phenomenon is a very exciting and interesting to investigate, that's why that problem can be proposed for the students' investigation at the high schools or a research project for training beyond the lessons. In spite of the fact that the phenomenon depends on a lot of parameters each of them can be investigated and described. It has been interesting to determine the influence of a regime of insonification on the state of the water surface. And next, to find a correlation between this state and the floating time of the droplets. It seems that the most important characteristics that may be used for study of the phenomenon are:

- sound frequency and corresponding wavelength,
- amplitude of surface waves,
- pattern of a surface mode as a function of amplitude and frequency.

4. Experimental part

Objects of investigation are low-concentration soap solutions in water, which have been placed in a Petri dish. Droplets which hover above the insonified water surface have been produced in two ways: (i) using a medical syringe and (ii) spontaneous droplets creation at a high level of sound.

Experimental setup

The sketch of an experimental setup is presented in Fig. 2.

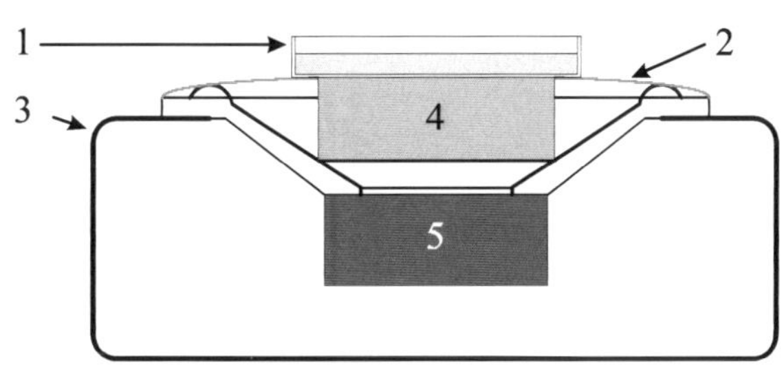

Fig. 2. Experimental setup. 1 – Petri dish with aqueous soap solution, 2 – kitchen scratch plastic tape, 3 – plastic box as a frame work, 4 – polyfoam pedestal, 5 – loudspeaker

It contains a mechanical part, self-made AC amplifier based on the TFA8943j chip, sound frequency generator as computer software. The relative amplitude of the electrical signal supplied to the loudspeaker was measured by a digital tester. Time of the droplets floating was registered by a stopwatch.

Surface peaks measurements

Subject to the actual magnitude of surface waves they have been measured using two methods: (i) side view photography for the large magnitudes, and (ii) laser grazing beam reflection for smaller ones. The scheme of the last method is presented in Fig. 3.

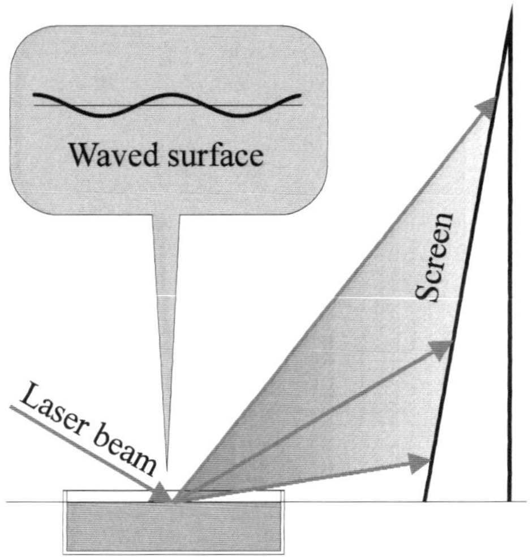

Fig. 3. Experimental setup for surface waves detection of a small amplitude using grazing beam laser probe

IOP Conf. Series: Journal of Physics: Conf. Series **1076** (2018) 012024 doi:10.1088/1742-6596/1076/1/012024

The obtained results allowed the determination of the relationship between measured values of relative voltage amplitude on the loudspeaker and frequency of the sound with the actual amplitudes of the surface waves.

5. Theoretical part

Model

In order to describe the effect of droplet floating it has been supposed that there is a thin air interlayer between the surface of the liquid (water in our case) and a droplet. Time needed for the droplet to coalescence to the main volume is the time when the distance between two surfaces exceeds a certain threshold. That time the air in the interlayer is squeezed out in the conditions of viscous flow. Equilibrium form of curved liquid surface under the droplet and form of the droplet itself are the topic of enough complicated calculations that may be implemented numerically only. Here we only estimate the influence of individual model parameters (geometry sizes and matter properties) on the lifetime of a droplet.

Usable geometrical parameters are presented in Fig. 4c. A curved liquid surface is modelled as a spherical indentation with deepness *h*, radius of edge circle is *R*, and spherical radius is R_{drop}. The last is also the radius of the droplet, which is believed to be spherical, the form distortions have been ignored.

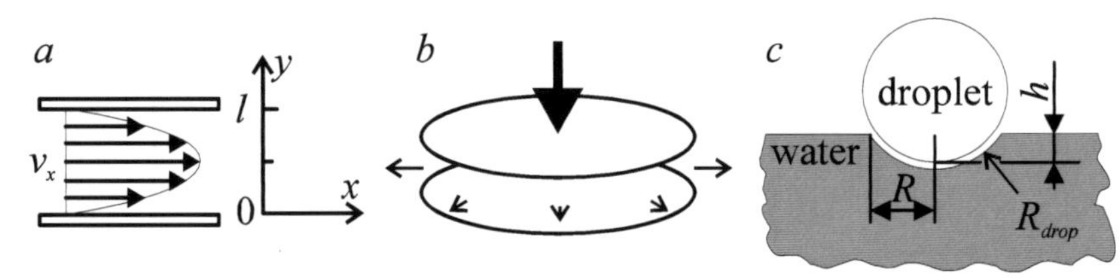

Fig. 4. Distribution of the velocities $v_x(y)$ in the viscose flow between two parallel plates distanced on *l* (a). Scheme of a viscous substance efflux from the space between parallel discs during the process of their closing (b). Schematic representation of a droplet hovering on the surface of liquid (c)

Theoretical estimations

At lamellar flow of viscous media along a space of a height *l* between two parallel planes in the direction *x* (see Fig. 3a) has been described by the following equation:

$$\frac{P(x)-P(x+\Delta x)}{\Delta x}=\mu\frac{8v_{max}}{l^2},$$

where *P*(*x*) is the normal pressure, and its loss along the direction of a flow has been expressed via the coefficient of dynamic viscosity μ and the maximal velocity of the flow in the center of the space v_{max}. The velocity distribution under those circumstances along a *vertical* direction *y* can be described by the equation:

$$v_x(y)=\frac{4v_{max}}{h^2}\left(yh-y^2\right),$$

which allows to determine an average velocity $v_{x,ave}=2v_{max}/3$.

In the case of a viscous media an efflux from the space between two closing discs, a radial velocity and the normal pressure distributions keep their functional binding. So we can rewrite the equation written above with polar coordinates:

IOP Conf. Series: Journal of Physics: Conf. Series **1076** (2018) 012024 doi:10.1088/1742-6596/1076/1/012024

$$\frac{\Delta P(r)}{\Delta r} = -\mu \frac{12\, v_{ave}(r)}{l^2}. \quad (1)$$

The continuity equation requires that a volume of the viscous medium flowing through the cylindrical lateral border has to be equal to the loss of the space between the disks, so $v_{ave}(r) = rU/2l$, where U is a velocity of the discs closing (Fig. 3b). Putting this equation into (1) and integrating it results in the explicit form of the radial pressure distribution function:

$$P(r) = P_{atm} + \mu \frac{3U}{l^3}\left(R^2 - r^2\right).$$

Because of the global character of P_{atm} we can exclude it from the further consideration keeping the part connected with the viscosity only:

$$\tilde{P}(r) = \mu \frac{3U}{l^3}\left(R^2 - r^2\right). \quad (2)$$

The force of the discs closing has been calculated by integration of (2) from 0 to radius of the disc R:

$$f_P = \mu \frac{3\pi U R^4}{2l^3}. \quad (3)$$

For the task of the droplet on the liquid surface f_P is its gravity force $\frac{4}{3}\pi R_{drop}^3 \rho g$. So, the expression for the quantity U may be easily obtained from (3):

$$U = \frac{8}{9\mu} \rho_{water}\, g l^3 \frac{R_{drop}^3}{R^4}. \quad (4)$$

The time of a motion t of the droplet while l exceeds some threshold value l_t has been calculated by integrating the reciprocal velocity U^{-1}, so called slowness. The result can be seen in the formula below:

$$t = \frac{9\mu}{8\rho_{water} g} \frac{R^4}{R_{drop}^3} \int_{l_t}^{\infty} l^{-3} dl = \frac{9\mu}{16 g l_t^2 \rho_{water}} \frac{R^4}{R_{drop}^3}. \quad (5)$$

For the estimation of *R* the principle of the potential energy minimum have been used. Potential energy of a droplet on the curved surface of liquid in frames of above mentioned model limitations (see Fig. 3c) has a form:

$$E = \pi \rho_{water}\, g \left(\frac{1}{12}\left(4R_{drop} - h\right) h^3 - \frac{4}{3} R_{drop}^3 h \right) + \pi \sigma h^2,$$

Where *h* is immersion depth of a droplet in the liquid, and σ is the coefficient of surface tension. Minimum is achieved at the conditions:

$$h^3 - 3R_{drop} h^2 - \frac{6\sigma}{\rho g} h + 4R_{drop}^3 = 0.$$

The value $r_0 = \sqrt{6\sigma/\rho g}$ can be interpreted as characteristic radius, which for the water at 20°C has value 6.68 mm ($\sigma = 72.9 \cdot 10^{-3}$ N/m, $\rho = 1000$ kg/m^3, $g = 9.81$ m/s^2). The quantity R included in (5) can be expressed by h and R_{drop} as $R = \sqrt{2hR_{drop} - h^2}$. It is convenient to represent above equation in dimensionless variable $\xi = h/R_{drop}$:

$$\xi^3 - 3\xi^2 - \left(r_0/R_{drop}\right)^2 \xi + 4 = 0. \qquad (6)$$

Its physically allowed root lies in the range 0-1. Analysis showed, that for $R_{drop} << r_0$ the root has asymptote $\xi = \left(2R_{drop}/r_0\right)^2$. That gives also the asymptotic behavior of R:

$$R \approx 2\sqrt{2}\, R_{drop}^2 / r_0 . \qquad (7)$$

Substitution (7) to (5) leads to the equation:

$$t = \frac{36\mu R_{drop}^5}{g l_t^2 \rho_{water} r_0^4}. \qquad (8)$$

Numerical calculation for the water droplets shows that the dependence $R\left(R_{drop}\right)$ is only slightly deviates from the asymptote (7) as can be seen in Fig. 5.

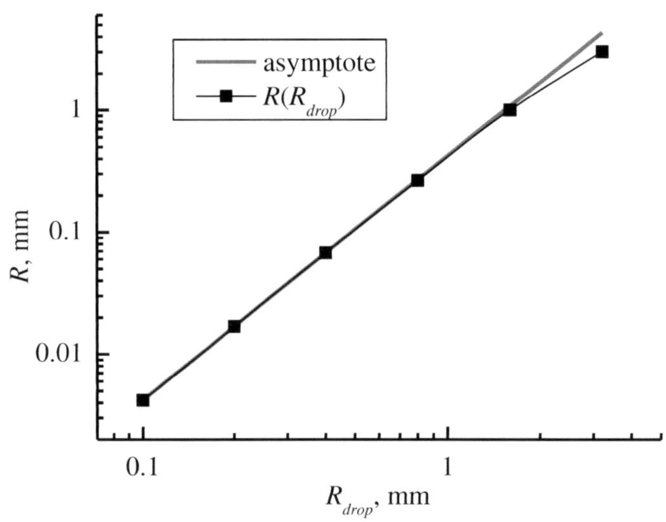

Fig. 5. The graphs of R by R_{drop}: squares – calculation using equation (6), red straight line is asymptote.
R is the radius of the contact circle between droplet of radius R_{drop} and the surface of liquid

6. Results and discussion

The analysis given above shows that the life time of the floating droplets depends on their radii essentially. Moreover the quantity r_0^4 in the denominator of (8) corresponds to the inverse square proportionality of a droplet lifetime to the coefficient of surface tension. Experiments were carried out mainly with one concentration of a soap solution. Experimental investigation of the surface tension influence on the behavior of this physical system is an open question.

The fact that the drop can hover a long time on the insonified surface of the liquid occurs due to frequent jumping, during which an interlayer of air insulating droplets from the surface is restored. It may occur only in the conditions when surface oscillation imposes acceleration to the droplet greater than the free fall acceleration. That is why we investigated parameters of waves on the liquid surface. Unfortunately surface waves depend in a complex way on intensity and frequency of the sound. Fig. 6 represents some examples of different surface wave modes. Such variety of behavior complicates investigations and requires examining of each type of surface mode.

Fig. 6. The photographs of different modes obtained at low 20 Hz (a, b) and high 150 Hz (c) frequencies of sound. The modes in (a) and (b) are produced at the same frequency, but different intensity of a sound. Pattern (a) appears at lower intensity

In the limit of low frequency and long wavelength droplet time of life is strictly dependent on its position. That is because the surface mode pattern contains areas with high amplitude of oscillation (loops) and small one (nodes). Our observations have shown that at favorable conditions (area of a loop) the lifetime of a bounced droplet can exceed several tens of minutes.

Photos of low-frequency mode patterns are shown in the Figs 6a and 6b. At the intensive insonification and high frequency the self-organized structure of nodes and loops on the surface of the liquid looks like a two-dimensional square lattice (Fig. 6c). Survival of a droplets under these conditions is caused by the fact that distances between loops becomes smaller than the droplet dimension, so in any position the droplet experiences action of the surface. Examples of the observed hovered droplets are presented in Fig. 7.

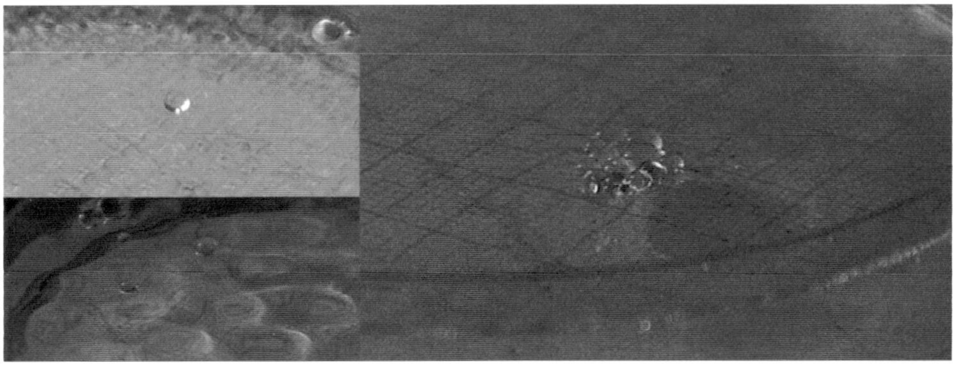

Fig. 7. The photographs of droplets that live a long time on the insonificated aqueous surface

Experimental observations and research project presented in this part of article are only the start of the investigation that will be continued in future. Of course the informal teaching techniques described in the article are not able to decide the problem of decreasing students' interest in physics. Only 15 students stayed at the third stage of preparation and usually nearly five of them throw their career with IT or enter the other universities on STEM specializations but every year the Faculty mention above bring up about 10 particularly motivated, with a strong interest in science and developed research competences who desire to connect their future life with research career.

Acknowledgement

The authors give lots of thanks to Prof. Dr. V. Lisovskiy and research engineer O. Gapon for the inspiring guidance, invaluably constructive advice and understanding during work with school children. We also give special gratitude to Dr. Vladymyr Farenik for the support and useful advice during the students' trainings.

References

Dvorak L. (2007) Labs outside labs miniprojects at a spring camp for future physics teachers. *European Journal of Physics* **28**, 95-104.

IYPT, http://iypt.org/images/e/ef/problems2016.pdf.

Kazachkova N. (2007) Creation The First in Ukraine Touch-Exhibition Of Physics Paradoxes As An Innovative Way Of Physics Popularization. In *Book of Abstract, GIREP-EPEC Conference Frontiers of Physics Education, Opatija, Croatia, 26-31 August, 2007*, pp. 150-151.

Kazachkova N., Yanson Y., Kryukov Y., Khodko A. (2000) Students Research Work Is One of the Innovative Methods of Physics Teaching. In *International Conference Physics Teacher Education Beyond 2000 and PTTIS, The Book of Abstract (27 August to 1 September). Barcelona – Spain, 2000*, p. 205.

Priemer B. (2006) Open Ended Experiments about Wind Energy. In *Book of Abstracts*, *GIREP Conference 2006 "Modelling in Physics and Physics Education* (Van den Berg E., Van den Berg D. & Ellermeijer T., eds.), Amsterdam, p. 77.

Sjoberg S. & Schreiner C. (2006) How Do Students Perceive Science and Technology? *Science in School* **1**, 66-68.

Trna J. (2005) Motivation and Hands-on Experiments. *Proceedings of the International Conference Hands-on Science in a Changing Education. HSci2005*, Rethymno: University of Crete, 169-174.

Low Cost 2D-Hover Puck

Josef Emanuel Pürmayr, Brigitte Wolny, Martin Hopf

University of Vienna

Abstract. The present contribution presents the development of a teaching tool for two-dimensional mechanics. A *Low Cost 2D-Hover Puck* has been designed and engineered, with which gravitational effects are locked out and motion in two dimensions can be shown. A horizontally aligned fan is mounted on top of the Hover Puck and provides constant acceleration. Now, experiments about motion with initial velocity and a constant acceleration in an arbitrary direction can be conducted by students or teachers. Due to the simple and inexpensive design, the Hover Puck can be assembled by students, by almost only using household objects and is therefore ideal for school.

1. Introduction

In traditional physics lessons, gravitational force is often used to teach constant accelerated motion. Free fall, an inclined plane or a horizontal throw, for instance, are typical settings for this teaching purpose. Gravity, however, as we know from literature (Stadler 1996), comes with a number of misconceptions and learning barriers for students (Müller *et al.* 2011, Wiesner *et al.* 2011).

> *Gravity is pervasive, acts invisibly, is not a subject to control, and is often invoked by students as a name without a well-defined concept behind it.* (Morse 1993)

As outlined by Morse, the two main learning barriers with respect to gravity are, that it acts invisibly and that it is not a subject to control.

Knowing the drawbacks of gravity as a source for acceleration, Morse (1993, 2005) developed a different approach towards teaching accelerated motion. Instead of the classical experiments, he introduced low-friction carts on rails, on a tabletop plane, to show constant accelerated motion, so that gravitational effects are locked out. In these carts acceleration is realized with one or more horizontally aligned fans. Using fans for acceleration brings a number of advantages over gravity. The accelerating effect no longer acts invisibly; it can be seen and even heard. Students even can roughly anticipate the thrust of two fan units, by comparing their pitch when operating (Morse 2005, p. 162). Contrary to traditional experiments, experiments with Morse's carts can manipulate and analyze mass, force and time of acceleration independently. Carts can be loaded with additional weights, or several carts (without fans) can be stacked onto each other to increase the total weight. The number of fans can be adapted, as well as their direction of thrust. Fans can be aligned to provide thrust in one direction, as well as they can be mounted onto the cart in an opposing way. These possible adjustments offer a range of possible experiments concerning one-dimensional accelerated motion, by simultaneously locking out gravitational effects (Morse 2005).

However, literature shows (Tobias 2010), that teaching motion in two dimensions is superior to teaching motion in a one-dimensional setting. Students show a significantly better understanding of the vectorial nature of motion, if a dynamical, two-dimensional approach towards mechanics is offered.

By creating a two-dimensional setting for teaching motion and simultaneously locking out gravitational effects, we can enhance Morse's carts to provide a beneficial learning environment for students in introductory mechanics.

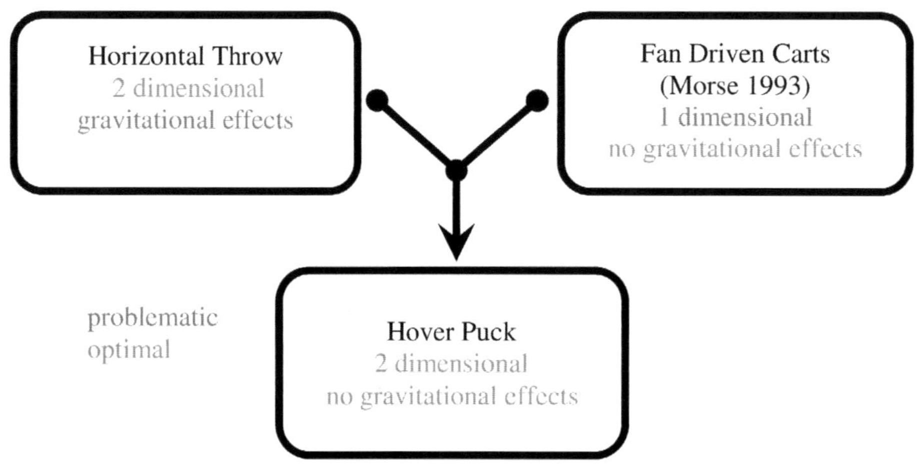

Fig. 1. Benefits of the Hover Puck

Fig. 1 shows the advantages and disadvantages of the traditional approach (Horizontal Throw) and Morse's (1993, 2005) more modern approach towards teaching motion with fan driven carts. Fig. 1 additionally depicts the idea of combining the advantages of two ways of teaching mechanics (throw and carts) to create a teaching material, which accommodates both, the exclusion of gravitational effects and a two-dimensional way of operation.

2. Beneficial learning environment

A two-dimensional curriculum for introducing motion

Building on an adapted concept of Jung, Reul and Schwedes (1977), Wiesner *et al.* (2011) developed a two-dimensional curriculum for introductory mechanics, which is already successfully implemented in Bavarian 7th grade (12 to 13 year olds). It introduces the concept of force in a dynamic way, by explicitly excluding the term *acceleration* from the equation of motion, resulting in:

$$\vec{F} \cdot \Delta t = \Delta\vec{v} \cdot m$$

By deliberately substituting acceleration by its integrative form $\vec{a} = \Delta\vec{v}/\Delta t$ and bringing the *time of interaction* to the *interacting force*, the concept uses the impulse formula as a basis for discussing the concept of accelerated motion. The focus of the mechanics curriculum lies on the vectorial nature of movement and encourages students to discuss changes in direction or speed based on a vector based understanding of motion.

A teaching tool for teaching motion in two-dimensions

A teaching tool for showing constant acceleration in a two-dimensional plane, can extend Wiesner *et al.'s* (2011) content structure in providing a simple hands-on-experiment for students to physically experience constant accelerated motion in two dimensions.

Concluding from these guidelines, a Low Cost 2D-Hover Puck for teaching motion in two dimensions has been designed and engineered.

Fig. 2. Hover Puck – view from the left (modified from Pürmayr 2017)

Fig. 3. Hover Puck – view from the rear (modified from Pürmayr)

One core point of designing the Hover Puck was to realize it as a low-cost experiment. Most of the components are household objects and the assembling can easily be done by students at home or at school.

For the wiring of the apparatus, a soldering kit is needed. The total costs of the Hover Puck sum up to about 15€, if propellers and DC-motors have to be bought.

The mount (base structure) for motors and batteries, as well as the propellers, can be 3D-printed by a data-file provided. It can be downloaded from the Website of the Austrian Educational Competence Centre for Physics[1].

If the setup and the propellers can be self-printed, the total costs of the Hover Puck are reduced to about 5€. Batteries are not included in the costs, since old cellphone-batteries are sufficient for powering the Hover Puck. They usually can be charged using an universal charger.

Fig. 4 shows the simple setup, as well as the color coded parts, which are itemized in the parts list in table 1 below. For reasons of simplicity the wires and garbage bag are not shown in Fig. 4.

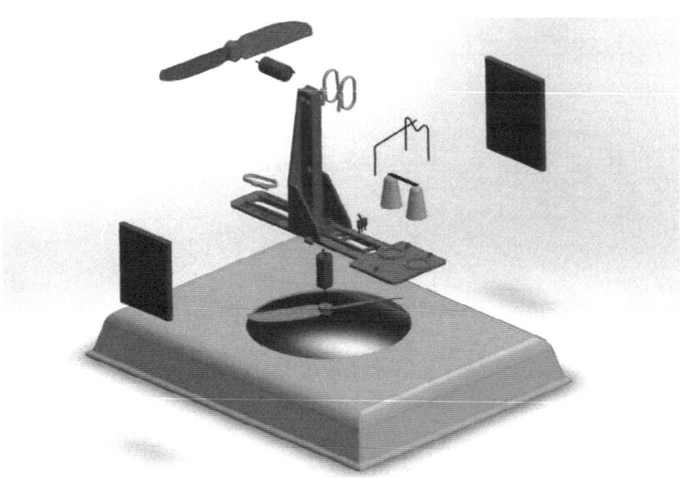

Fig. 4. Construction view of the Hover Puck (Pürmayr 2017, p. 58)

[1] http://aeccp.univie.ac.at/fileadmin/user_upload/kompetenzzentrum_aeccp/Materialen_Physik_Verstaendlich/Mechanik/Hover Puck.zip.

Table 1. Parts of the Hover Puck. The color coded parts from Fig. 4 are labeled in the column *Color Code* (modified from Pürmayr 2017)

Component	Quantity/ Length	Cost per Piece [€]	Notes	Color Code
Styrofoam tray	1	0	Household object	yellow
Zipties	3	0	Household object	yellow
Foam ear plugs	2	0	Household object	yellow
Duct tape	1 m	0	Household object	yellow
Garbage bag	1	0	Household object	yellow
Paper clip	3	0	Household object	yellow
Base structure	1	0.25	3D-print	green
Vertical fan	1	0.05	3D-print	green
Horizontal fan	1	0.05	3D-print	green
Motors	2	0.73	Part to be purchased	red
Miniature switch	1	3.09	Part to be purchased	red
Wire	0.4 m	0,07	Part to be purchased	red
Cellphone batteries (SAMSUNG)	2	0	Picked up from recycling facility	violet
TOTAL COSTS		4.97		

We chose a three-way miniature switch to provide possibilities for expanding the current modes of operation of the Hover Puck (E.g. adding a different thrust of the horizontal fan).

Mechanics of the Hover Puck

Fig. 5 shows a simplified version of the Hover Puck in a sectional view. Two fans (one vertical and one horizontal) are used to power the Hover Puck.

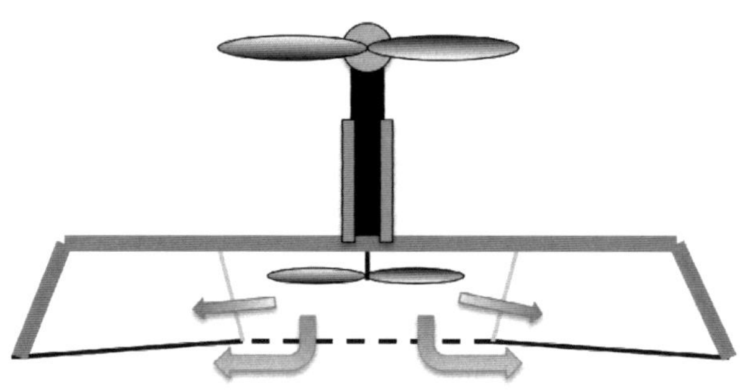

Fig. 5. Sectional view of the Hover Puck (modified from Pürmayr 2017)

Used phone-batteries provide electrical power for the motors and fans and are placed right next to the vertical structure, which serves as a mount for the horizontal motor. The placement of the batteries (Fig. 5, blue bars) has a crucial impact on a stable run of the Hover Puck, since a centered distribution of mass is required for balancing the Hover Puck.

A styrofoam tray (Fig. 5, red) is turned upside down and serves as a base. The open side of the tray is covered with a plastic bag (Fig. 5, black), which is lifted up by two threads (Fig. 5, grey). A round cutout (dashed line), right under the vertical fan, allows air flowing underneath the Hover Puck (blue arrows). The green arrows (Fig. 5) represent the portion of the air, flowing sideways and blowing up the plastic foil. Due to the increased pressure inside the styrofoam tray, the puffed up plastic foil compensates uneven patches such as joints or small gaps in the underground.

IOP Conf. Series: Journal of Physics: Conf. Series **1076** (2018) 012025 doi:10.1088/1742-6596/1076/1/012025

Due to the air flow to the sides (Fig. 5, green) and under the Hover Puck (Fig. 5, blue), an air cushion forms underneath the Hover Puck, reducing friction to a minimum. The residual friction results from sliding friction between the foil and the surface material.

Now, motion in two dimensions on a tabletop plane can be shown and can even be expanded by a component of constant acceleration in an arbitrary direction. The horizontal acceleration of the Hover Puck is realized by the horizontally aligned fan (Fig. 5).

The fan can be activated manually, or by a time delay switch, allowing creation of a non-influenced setup for experiments.

Fig. 6 shows the mechanics of the time delay switch. Two blank paper clips (blue and red) are bent, serving as contacts. The wires leading to the battery are represented by the black curved lines and are soldered to the bent paper clips. In its open position, the bent paper clips don't connect to each other and therefore, the electric circuit is open. As shown in Fig. 6 on the right, foam earplugs are used to activate the horizontal fan, by pushing the blue wire up against the red wire, closing the circuit. Another paperclip (Fig. 6, black) is used as a web, connecting the foam ear plugs. The time of activation can be delayed from 1 to 20 seconds, by the amount of force applied to press down the ear plugs.

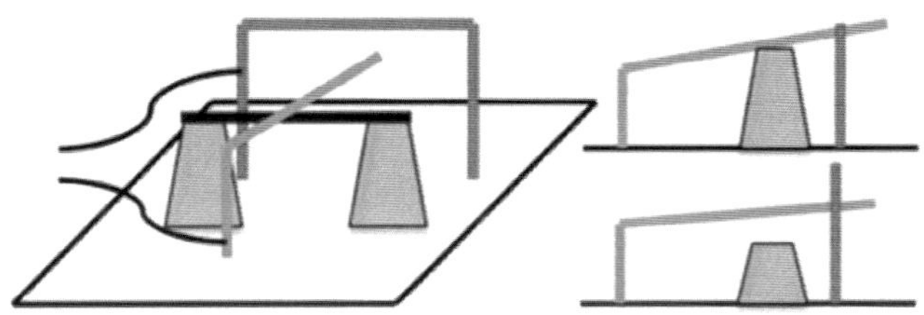

Fig. 6. Time delay switch (Pürmayr 2017)

For an adequate use of the Hover Puck for introducing motion in a two-dimensional setting, the constancy of acceleration is key. The Hover Puck has been tested for the change of acceleration over time to ensure its suitability for a two-dimensional approach towards mechanics.

Fig. 7 shows the linear acceleration of the Hover Puck over time. The x-axis shows the time in seconds and the y-axis depicts the acceleration of the Hover Puck. The unit of the acceleration in Fig. 7 is arbitrary, for the magnitude of the acceleration highly depends on the conditions of the ground's surface.

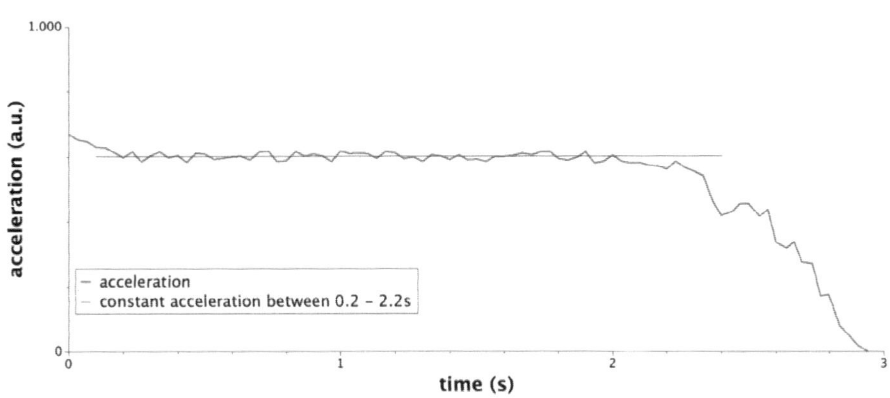

Fig. 7. Constant acceleration over time of the Hover Puck (Pürmayr 2017)

As Fig. 7 reveals, acceleration can be considered as constant between 0.2 and 2.2 seconds, creating the possibility to teach constant accelerated motion in a two-dimensional setting.

Examples for viable Experiments

Figs 8 and 10 show two examples for possible experiments with the Hover Puck.

Two-dimensional motion with an initial velocity to the left and constant acceleration in a forward direction can be shown. The Hover Puck is set in motion in a direction perpendicular to the acceleration of the Hover Puck (Fig. 8). The horizontal fan is activated right at the start, showing, in combination with the initial velocity to the left, a constantly accelerated motion.

The vectors of the velocity to the left (red) and the velocity forward (green) add together to the resulting velocity (blue) of the Hover Puck at a certain time. The construction of the vectors of velocity can be done by students, to provide an insight into the mechanics of motion. Fig. 8 has been created using a free motion shot app on a smartphone.

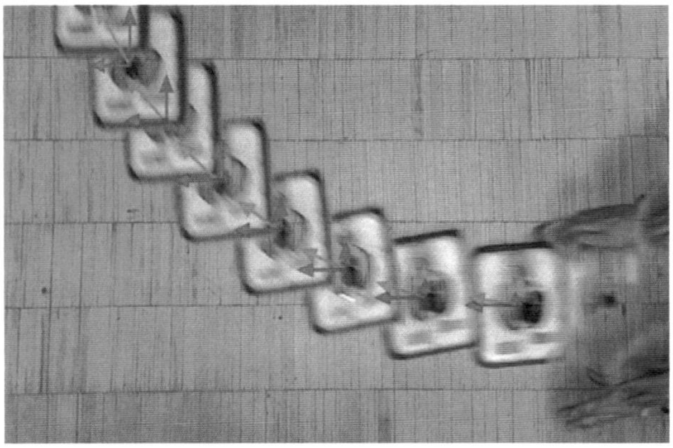

Fig. 8. Constant acceleration perpendicular to initial velocity to the left (Pürmayr 2017)

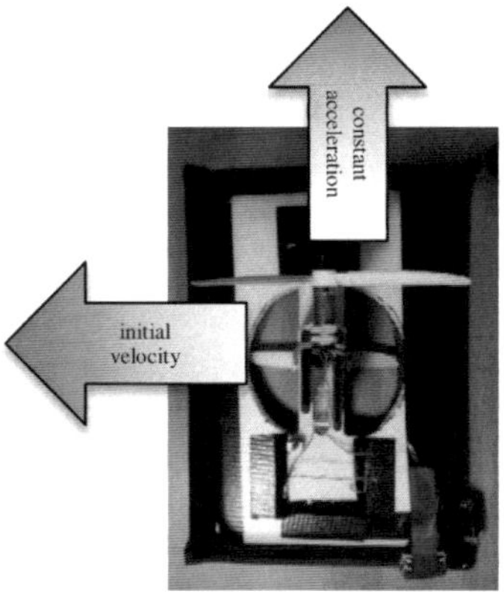

Fig. 9. Directions of initial velocity and acceleration (Pürmayr 2017)

The Hover Puck can also be used to show one-dimensional accelerated motion (Fig. 10). The time-delay switch provides a non-manipulated system by delay triggering the activation of the horizontal motor.

Just as in the previous example, the stop-motion illustration has been created with a smartphone-app to show the acceleration.

Fig. 10. Linear constant acceleration of the Hover Puck (Pürmayr 2017)

3. Outlook

For further development of the Hover Puck, Austrian teachers, with experience with Wiesner *et al.'s* (2011) mechanic curriculum have been interviewed. The goal was to gather information about possible applications and difficulties concerning the Hover Puck in Physics classes. This feedback will be used to improve the Hover Puck to become a viable teaching tool for teaching introductory mechanics, which will be tested in school for its effectiveness on learning.

References

Jung W., Reul H. & Schwedes S. (1977) *Untersuchungen zur Einführung in die Mechanik in den Klassen 3-6.* Frankfurt am Main: Diesterweg.

Morse R.A. (1993) Constant Acceleration: Experiments with a Fan-Driven Dynamics Cart. *The Physics Teacher* **31**(7), 436-438.

Morse R.A. (2005) Fan Unit Physics. *The Physics Teacher*, 43, 162. https://doi.org/10.1119/1.1869427.

Müller R., Wiesner H., Hopf M. & Wodzinski R. (eds.) (2011) *Schülervorstellungen in der Physik: Festschrift für Hartmut Wiesner* (3rd ed.), Köln: Aulis Verlag, p. 108.

Pürmayr J.E. (2017) *Entwicklung und Evaluierung eines Luftkissengleiters für einen zweidimensionalen Mechanikunterricht*, University of Vienna, Vienna, pp. 43, 58-67.

Stadler H. (1996) Pupils' Conceptions about Gravity. *GIREP Conference Proceedings, 1996*, 378-380.

Tobias V. (2010) *Newton'sche Mechanik im Anfangsunterricht: die Wirksamkeit einer Einführung über die zweidimensionale Dynamik auf das Lehren und Lernen*. Berlin: Logos-Verlag.

Wiesner H., Schecker H. & Hopf M. (2011) *Physikdidaktik kompakt*, Hallbergmoos: Aulis Verlag, p. 38.

Wiesner H., Wilhelm T., Waltner C., Tobias V., Rachel A., Hopf M. & Schwarze H. (2011) *Mechanik 1: Kraft und Geschwindigkeitsänderung*. Hallbergmoos: Aulis-Verlag.

IOP Conf. Series: Journal of Physics: Conf. Series **1076** (2018) 012026 doi:10.1088/1742-6596/1076/1/012026

Arduino as a tool for physics experiments

Giovanni Organtini

Sapienza Università di Roma & INFN Sez. di Roma

Abstract. Arduino is a widely used open-source platform composed of both hardware and software tools that can be very useful in a physics laboratory. Its low price, the large availability of sensors and transducers, its ease of use and its open nature makes it a perfect tool to involve students in active and cooperative learning. In this paper we show a few examples of what can be done using an Arduino and some sensor, and propose a template for documenting activities.

1. Introduction

New trends are emerging in education and, in particular, in physics education. Besides inquiry based learning (IBL), cooperative learning (Slavin 2011) and project based learning (Krajcik & Blumenfeld 2006) are being recognised as effective learning strategies. Moreover, the integration of physics with other disciplines, in the past, was very limited to integration with mathematics. Nowadays, the ability of integrating different disciplines in learning strategies is considered very important. In particular, the need for the development of some ability in coding and in so-called tinkering, is widely accepted, and integration with various forms of art is becoming to be advisable. Already in 1988 Italo Calvino wrote that "it is true that software cannot exercise its powers of lightness except through the weight of hardware. But it is the software that gives the orders" (Calvino 1988): a very clear tribute to the fact that the ability to organise and manipulate data is much more important than the tools used to do that.

Teaching physics should not be limited to the transfer of physics laws to pupils: the reason for which we should teach physics to our students in schools is that physics is a very successful example of scientific methodology. We should then teach them how to apply methodology in physics, in order to develop their abilities in applying similar strategies in everyday life, rather than expect from them to be able to solve problems that, in most cases, they will never face.

With this respect, it is somewhat bizarre that pupils are practically trained in many disciplines, but not in scientific ones. Only exceptional students learn how to do physics: they usually learn the physics produced by others, not the physics made by themselves, nor how to do that. Learning physics this way is like learning literature without doing a composition, or learning art without doing a drawing, or even learning music without having tried to play an instrument.

A very common reason for this is that doing physics, at least until recently, needs a laboratory equipped with many expensive and complicated instruments. Where available, instruments are often limited in number and cannot be used at the same time by each student. Most of the time a teacher or a technician uses the available equipment to show some effect to the pupils.

With the advent of Arduino (www.adruino.cc), it is possible to fully equip a laboratory with enough instruments to allow each student to perform a lot of experiments and measurements with limited investment. Moreover, learning the Arduino basics requires the development of some coding ability and the open nature of the product promotes cooperative learning. Project based learning is a natural consequence of the need to design and build the experiment with the provided material.

In this paper we illustrate how Arduino can be a useful tool for a physics teacher, assuming that the reader is completely unfamiliar with it. In the following sections we describe the Arduino platform and the most important programming tools for using it as a tool for doing physics, together with examples of experiments.

2. The Arduino platform

Arduino is an open source platform composed of electronic boards, sensors and expansion boards, as well as a software development environment, initially developed in Italy by a team lead by Massimo Banzi. Thanks to its low cost and the free availability of its hardware design and software, the Arduino platform rapidly became an international standard and is now widely adopted all over the world for various kinds of projects: from fast prototyping to Internet of Things (IoT) projects.

The main boards exist in various flavors and form factors. The simplest and most widely used one is the Arduino UNO, based on the ATmega328P chip running with a clock at 16 MHz. All the boards share the same programming environment and may differ for form factor, memory size, number and type of ports and speed. For the purpose of a physics laboratory the Arduino UNO board is perfectly suitable. The board can be bought for as low as about 20 euros.

Fig. 1. The Arduino UNO board. On top of the figure there is a row with 14 digital pins. Power and analog pins are on the opposite side

The Arduino UNO board (Fig. 1) has 14 digital I/O pins. Digital pins may have two states: LOW and HIGH, corresponding to a voltage of 0V and 5V, respectively. When used as output ports, digital pins can provide a current up to 20 mA.

Six out of 14 pins can be used as Pulse Width Modulation (PWM) outputs. They can be assigned a value between 0 and 255 and, correspondingly, the pin stays HIGH for a percentage of time between 0 and 100% with a 2 ms base.

It also has six analog inputs, each with ten bits resolution, and with a dynamic range from 0 V to 5 V. Power ports provide a 5 V and 3.3 V source together with the ground reference (GND) and provides currents up to 50 mA.

IOP Conf. Series: Journal of Physics: Conf. Series **1076** (2018) 012026 doi:10.1088/1742-6596/1076/1/012026

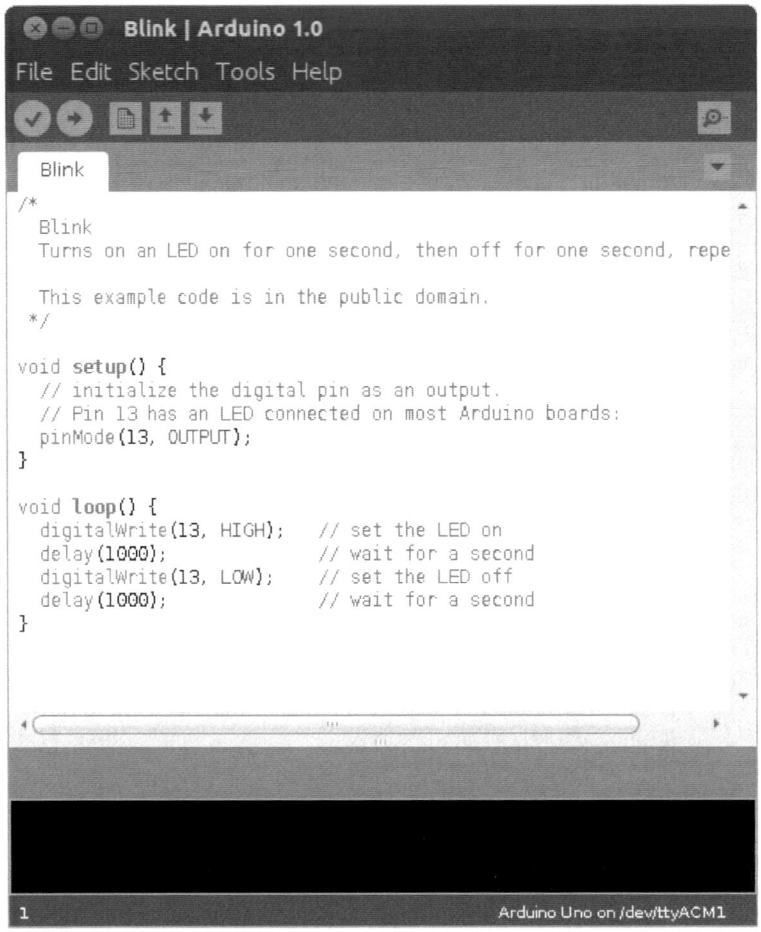

Fig. 2. The Arduino IDE window. The program is written inside the main, white window. The program is transferred to the board using the button with an arrow on the top left side of the window

A USB connection allows a user to program the chip using the Integrated Development Environment (IDE) software, freely available from the Arduino website. The user write his/her own program on a PC, compiles it and then transfers the program to the Arduino memory, that starts immediately executing it every time it is powered on. Arduino boards can be powered either using the USB cable connecting it to the PC for programming it, or via an external source providing from 7 to 12 V (a 9 V battery is perfectly suitable).

3. Arduino programming

The board is fully programmable in C++. The programming environment is such that a deep knowledge of Object Oriented Programming is not needed. Contrary to a computer, a microprocessor does not run an operating system: it just executes the only task loaded in its memory. In order to write the program, the user can use the freely available IDE on a PC. When launched it appears as in Fig. 2.

The program is entered in the main, white window as in most common text editors. Once ready, the syntax of the program can be checked using the checkmark button and, if correct, can be transferred to the board acting on the button with an arrow seen on the top left part of $\vec{F} \cdot \Delta t = \Delta \vec{v} \cdot m$. Once transferred, the program starts soon and restarts from scratch each time the board is powered on. Programming an Arduino consists of writing a sequence of operations that the microprocessor has to perform once, at the beginning of the run, collected under the function `setup()` (see Fig. 2), and a sequence of operations that the microprocessor executes repeatedly, until it is powered off, within the a function called `loop()`.

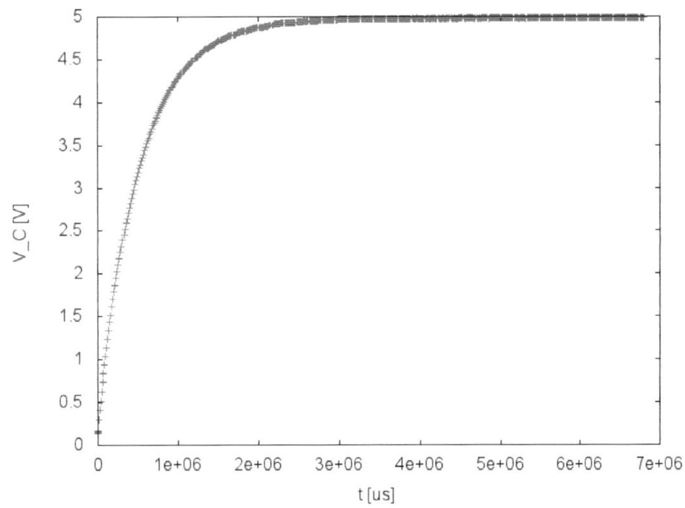

Fig. 3. The voltage across a capacitor in an RC circuit as measured by a Arduino UNO board connected to an RC circuit

Programmers may add functions to the program, even if they are not strictly needed. They can be regarded just as a way to keep the code well organised and simple enough. It must be noted that, in order to perform most of the experiments described below, no particular programming abilities are required and that, in most cases, the needed abilities can be gained in one to three days of trial and error sessions even in the absence of formal training sessions.

We describe the main software functions for using Arduino together with the relevant examples in the sections below.

Using analog ports

A very basic usage of analog ports consists in performing simple electrical measurements. Analog ports can be electrically connected to any voltage source from 0 V to 5 V. The Arduino Analog to Digital Converter (ADC) converts the input voltage into a ten bits number from 0 to 1023. Hence, connecting any part of a circuit to an analog input pin, with the ground in common with the Arduino GND pin, allows one to perform electrical measurements with a resolution of 5 V/1024 $\approx$ 5 mV.

A great advantage with respect to the usage of a common multimeter is that the measurements can be taken repeatedly at short intervals and/or based on a given pattern. For example, a RC circuit can be easily realised using commonly available capacitors and resistors without the need for choosing large values for both C and R to let the student measure the voltage across the capacitor during the charge and the discharge phase.

Thanks to the 16 MHz clock, measurements can be repeated many tens of times every microsecond, allowing very detailed and accurate measurements of the voltage V_C across the capacitor versus time, as shown in Fig. 3.

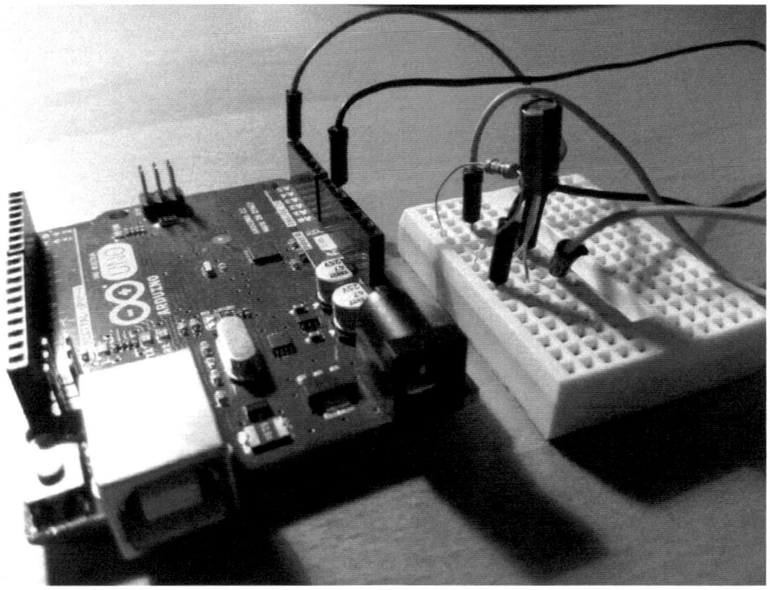

Fig. 4. An Arduino board connected to an RC circuit using jumper wires. The circuit is realised with the help of a breadboard

Building the circuit, in this case, is extremely simple. It is enough to connect a terminal of the capacitor to the GND pin, the resistor in series to the capacitor and the free lead of the resistor to the 5 V pin of the Arduino, while keeping the other terminal of the capacitor connected to one of the analog pins of the Arduino board. Connections can be easily done using a breadboard and jumper wires, as seen in Fig. 4.

Programming an Arduino to perform this kind of measurement is straightforward. The `setup()` function is as simple as

```
void setup() {
 Serial.begin(9600);
}
```

while the loop() function reads as

```
void loop() {
  Serial.println(analogRead(A2)*5./1023);
  delay(150);
}
```

The `Serial.begin(9600)` statement (statements are separated by a semicolon) lets Arduino connect to the PC via the USB port at a speed of 9600 bauds (a unit of measure of the transmission speed). The `loop()` function consists of a statement to convert the voltage at the analog pin A2 in a ten digit number (`analogRead(A2)`) and to convert it to Volts using the conversion factor 5/1023 as explained above. The result of the operation is sent to the PC via the USB connection using the `Serial.println()` statement. The `delay(150)` statement causes the microprocessor to wait for 150 ms before executing again the loop.

This extremely simple program performs a measurement of the voltage between the GND and the A2 pin every 150 ms and prints the values read on the “serial monitor”: a tool available in the Arduino

IDE consisting in a window to which each `Serial.println()` statement causes the appearance of the number inside the parenthesis. Each time the statement is invoked a new line is written.

The Arduino IDE also provides a "serial plotter": a window to which a graph appears based on the numbers sent to the PC via the statement above.

Using digital ports

Digital ports can be used in a variety of ways. For example, with ultrasonic sensors, one can perform interesting measurements in the field of mechanics. Position (in one dimension) versus time can be measured using available modules made of two piezoelectric devices (like, e.g. the HC-SR04): one used as a transmitter and the other as a receiver.

The module has four pins: two of them (GND and VCC) are used to power it and must just be connected to the corresponding GND and 5 V Arduino pins. The other two pins are called "trigger" (TRG) and "echo" (ECHO). Sending a square wave of at least 10 μs width to the TRG pin causes the module to produce a train of ultrasound waves from the transmitter. If the waves encounter and obstacle, they are reflected back and can then be detected by the receiver, that raise the ECHO pin from LOW to HIGH. Measuring the time t elapsed from the trigger to the echo, one can obtain the distance d of the obstacle knowing the speed of sound *c* in air, as d = ct/2.

A set of repeated measurement of distances and times can be obtained with the following code (in the example the TRG pin is connected to Arduino digital pin 2, while the ECHO pin is connected to the Arduino digital pin 3; row numbers are added for reference in the text):

```
 1. void loop() {
 2.    int x = 0;
 3.    digitalWrite(2, HIGH);
 4.    delayMicroseconds(10);
 5.    digitalWrite(2, LOW);
 6.    x = C*pulseIn(3, HIGH)/2.;
 7.    t = micros();
 8.    Serial.print(t);
 9.    Serial.print(" ");
10.    Serial.println(x);
11. }
```

The `digitalWrite()` function allows a programmer to set the state of the given pin to one of the possible states of the pin (LOW or HIGH). The `delayMicroseconds()` function put the microprocessor in pause for the number of microseconds specified in parenthesis. Rows from 3 to 5 produce a square wave as requested by the ultrasonic module: digital pin 2 is initially in the LOW state; it is then put at the HIGH state and back to LOW after 10 μs.

The `pulseIn()` function takes two arguments: a pin and a state. It waits until the given pin (the ECHO pin in this case) changes into the specified state (in the example from LOW to HIGH) and returns the time elapsed in μs. The integer variable x is then assigned with the distance measured as outlined above using the constant C defined elsewhere as the speed of sound m/μs. The t variable, instead, defined elsewhere, is set to the current time using `micros()`, a function returning the elapsed time since the beginning of the program in microseconds. Lines 8 to 10 print the values of t and x on the serial monitor.

This is just an example of the enormous number of possible measurements that can be done using the digital pins. Indeed, digital pins can be used to represent data in binary form and can then be used for much more complex measurements. A variety of dedicated modules exists engineered such that the values measured by onboard sensors are digitised on the module itself and transmitted to Arduino using the I2C protocol via digital pins.

4. External transducers

The open nature of the Arduino platform stimulated the production of many, cheap expansion boards, called "shields", that just plug onto the Arduino board and provide specialised functions. Interface shields and transducers of various types can be bought for as low as 1-2 euros and usually do not exceed 10 euros each, depending on the complexity of the systems.

Sensors for temperature, humidity, pressure, light, sound, acceleration, magnetic field, currents, etc. exist, allowing the realisation of lot of physics experiments, ranging from calorimetry to electromagnetism, from thermodynamics to the physics of waves. The only limit is the ability in designing an experiment that, in turn, can be developed with time and experience gained on the simplest ones.

Using the Arduino platform to perform the measurements has another advantage with respect to specifically designed equipment: the latter appears to students more like "black boxes". The student does not really understand how they work and just get results from them. Using Arduino, students can understand much more deeply the principles of operation of the system. Moreover, using Arduino allows to appreciate many aspects of the measurement process that are usually neglected using dedicated, "closed" apparatus. As an example, consider the measurement of the distance of an object using an ultrasonic sensor: the speed of sound, e.g., depends on temperature, pressure and humidity, at least. The size of the obstacle matters, as well as the presence of other sources of reflections around the experimental area. A proper, detailed analysis of these sources of systematic error can lead to interesting, instructive discussions.

5. Resources

There are plenty of tutorials on Internet about how to operate Arduino. Even if they are very instructive, they are not specifically designed for building scientific experiments. Only recently the pedagogical content of the Arduino platform is being recognised and we expect the number of dedicated resources to increase in the future.

We already started to provide some dedicated resources: we are going to organise regular sessions of a school of physics with Arduino, to take place at least twice per year, and we made available an e-book about the scientific use of Arduino on our website (Organtini 2017). During the school of physics with Arduino we ask participants to design and implement physics experiments using such a technology. A brief documentation of each experiment is then made available through our website (http://www.phys.uniroma1.it/fisica/Arduino-Smartphone-Esperimenti). The documentation is still only available in Italian, but we plan to translate it in English, too.

We look forward for collaborators willing to help in the realisation of detailed guidelines about specific experiments that can be done with Arduino and to create an international network of experts to organise regular international workshops on this subject.

References

Calvino I. (1988) *Six Memos for the Next Millennium*. Cambridge, USA: Harvard University Press.

Krajcik J.S. & Blumenfeld P.C (2006) Project Based Learning. In *The Cambridge Handbook of the Learning Sciences* (Sawyer R.K., ed.), New York, USA: Cambridge University Press.

Organtini G. (2017) *Scientific Arduino Programming*. Available for download at http://www.roma1.infn.it/people/organtini/publications/scientificArduino.pdf.

Slavin R. (2011) Instruction Based on Cooperative Learning. In *Handbook of research on learning and instruction* (Mayer R.E. & Alexander P.A., eds.), New York, USA: Routledge.

Institute of Physics
Dirac House, Temple Back
Bristol BS1 6BE UK

ISSN: 1742-6588
ISBN 978-1-5108-7511-1
9 781510 875111

CURRAN ASSOCIATES INC.
proceedings.com